BEEKEEPING
Some tools for agriculture

Introduced by Eva Crane

Intermediate Technology Publications 1987

Practical Action Publishing Ltd
27a Albert Street, Rugby, CV21 2SG, Warwickshire, UK
www.practicalactionpublishing.org

First published 1987
Printed on demand, 2020

ISBN 9780946688883

A catalogue record for this book is available from the British Library.

Since 1974, Practical Action Publishing has published and disseminated books and information in support of international development work throughout the world. Practical Action Publishing is a trading name of Practical Action Publishing Ltd (Company Reg. No. 1159018), the wholly owned publishing company of Practical Action. Practical Action Publishing trades only in support of its parent charity objectives and any profits are covenanted back to Practical Action (Charity Reg. No. 247257, Group VAT Registration No.
880 9924 76).

Contents

Page

HOW TO USE THIS GUIDE

This is one of the twelve sections of the latest edition of *Tools for Agriculture*. Each section is introduced by a specialist who sets the range of tools, implements and machinery available against the background of good farming practice, and the factors best considered when making a choice.

It is intended the guide should be used by the following categories of people:

- Farmers' representatives who purchase equipment on behalf of their clients;
- Advisers who seek to assist farmers and farmers' organizations with the purchase of equipment;
- Development Agency personnel who need to purchase equipment on behalf of farmers and farmers' organizations;
- Prospective manufacturers or manufacturers' agents who wish to have information on the range of equipment currently available.
- Development workers, students and others who wish to learn about the types of equipment available.

We expect the reader to use the guide in one of the following ways:

- to find the name and address of the manufacturer of a specific piece of equipment whose generic name is known e.g. a treadle-operated rice thresher or an animal-drawn turnwrest plough.
- to find the name and address of the manufacturer of a piece of equipment whose general purpose is known e.g. a machine for threshing rice or soil tillage equipment.
- to find out about specific types of equipment or equipment used for specific purposes.
- to find out about the range of equipment available from specific manufacturers or manufacturers in particular countries.
- to learn more about equipment used for the different aspects of crop and livestock production and processing.

Guidance across the broad range of small-scale farming equipment is available, with indexes and cross references, in the parent volume.

Within each section the information is presented in three ways:

- A clear *introduction* which lays out the most important points to bear in mind when purchasing a particular type of equipment. (The emphases vary from section to section — showing the difficulty of decision-making when selecting equipment for smallholder agriculture.)
- *Comprehensive Tables* which list the manufacturers of certain types of equipment and give some further information about specific items, or the range of items manufactured. In many tables it was impossible to give the full address of the manufacturer and the reader is referred for these to the Manufacturer's Index.
- Pages laid out in grid pattern in which the compiler has attempted to present the equipment in a logical order, that in which operations are carried out, and within each type of operation the progress is from hand-operated, through animal-drawn, to motorized equipment. Sometimes one particular type of equipment is illustrated to represent a group — many of which may differ in detail, though not in use. Wherever possible the trade name of the equipment is used, in order to facilitate enquiries to the manufacturers.

Having located a manufacturer for the type of equipment in which you are interested, we suggest you write direct to the manufacturer for further details: current prices, availability, delivery times and so on. (Remember that, where known, telephone and telex numbers have been included in the manufacturers' index).

Every attempt has been made to ensure accuracy of the details presented in this guide, but doubtless changes will have occurred about which the compilers are unaware. We apologize to any reader to whom we may have given a false lead. A note will be made of up-to-date information which becomes available to ITDG.

It must be stressed that this guide relies on information supplied by the manufacturers and that inclusion of an item is no guarantee of performance. Whilst every care has been taken to ensure the accuracy of the data in this guide, the publishers and compilers cannot accept responsibility for any errors which may have occured. In this connection it should be noted that specifications are subject to change without notice and should be confirmed when making enquiries and placing orders with suppliers.

GENERAL INTRODUCTION

Problems of farmers in developing countries

The main economic characteristic of agriculture in developing countries is the low level of productivity compared with what is technically possible. It has been shown in many and varied circumstances that although farmers may be rational and intelligent, technological stagnation or slow improvements can still be the norm. This contradiction can be explained by understanding several unusual, troublesome features of agriculture. First, because agriculture is basically a biological process, it is subject to the various unique risks of weather, pests and disease which can affect the product supply in an unpredictable fashion. Despite exceptional biological risks, most farmers nowadays rely to various extents upon cash derived from sales of produce. But agricultural products have consumer demand patterns which can turn even good production years — when biological constraints are conquered — into glut years and therefore financial disasters. The biological nature of production also results in a large time-gap, often months or even years, between the expenditure of effort or cash and the returns. Once cash inputs are used, an unusually high proportion of working capital is required, compared with industry. The final problems created by the biological nature of production lie in the marked seasonality. The peaks of labour input create management problems, and perishable commodities are produced intermittently; both create additional financial and technical storage problems.

A second characteristic of agriculture is from the small scale of most farming operations, often coupled with a low standard of education of the operators, which gives farmers little economic power as individuals and little aptitude to seek such remedial measures as do exist. There are many examples of appropriate technology but small farmers will often need intermediaries, such as extension workers and project personnel, to open their eyes to the potentialities. Given the vulnerability of small farmers to biological and economic risks, those intermediaries have special responsibility to assess the impact of any new technology for each particular set of local circumstances.

A third factor which affects efficiency in agriculture is a political one. It is in some ways ironic that in countries with very large numbers of small farmers, producers tend to command little political power despite their combined voting strength. Indeed they are often seen as the group to be directly and indirectly taxed to support other, generally urban-based, state activities. As a contrast, in rich countries, we often see minorities of farmers with little voting power receiving massive state subsidies, much of which supports technological advancement. The rationale of farmers referred to above thus leads to the exploited, small farmers producing well below potential and the rich, large-scale farmers producing food mountains that can only be sold at further subsidized prices.

The 1970s food crisis, the recent failure of agriculture to match rising food demands in many countries, particularly in sub-Sahara Africa, and the failure of industry to fulfil its promise of creating employment and wealth has turned the attention of policy-makers back to the long neglected and often despised agriculture sector. New technology for the large number of low-income, small-scale, poorly educated farmers will be necessary if agriculture's enhanced role is to be successful.

What are the technology options?

Innovation and technology change has been and will be the main engine of agricultural development. Technology

Innovative equipment can be simple in construction: a four-furrow row seeder.

change can be described as the growth of 'know how' (and research as 'know why'). But technology is not just a system of knowledge which can be applied to various elements of agricultural or other production to improve levels or efficiency of output. Technology application requires and uses new inputs. In contrast, technique improvement is the more difficult art of improving production essentially with existing resources. Pity the poor agricultural extension worker sent out to advise experienced farmers with no new technology, but only improvements in technique to demonstrate!

It is possible to exaggerate the lack of prospect for improvement and the consequent need for new investment in farm resource use. Changes both on and off-farm are influencing the economies of traditional systems. For example, with farm size halving every twenty years or so in some regions — as a result of population growth, and with increasing demand for cash from farming activities for production items such as seed (which used to be farm produced) or for consumption items such as radio batteries and so forth, there are new challenges to the traditional rationale and the old system optima. But despite the need to adjust the existing resources to find new optima, the opportunities for really big gains will undoubtedly come from new technology which will often require radically different ways of doing things. A change in the resource base or the injection of a new piece of technology into an interdependent agricultural system may alter various other constraints and opportunities within the overall farm system. One function of a reference book such as this is to act as an encyclopaedia illustrating alternative ways of coping with new challenges. Readers do not have to reinvent the wheel each time a new transport system emerges. Self-reliance has little merit over technology transfer when it comes to solving food availability problems in a rapidly changing world.

This book displays a very wide range of technology and describes both what the technology can achieve, and how and where most information can be discovered. It shows that there is already in existence a mass of tested technology for small-scale farmers. The farm technology itself is laden with opportunities for improving the returns to land, water, labour and other crucial resources. The careful farmer, with help, can have many options.

The role of information

In the theory of classical economics, information on the contents of the technology itself is assumed to be a free good, readily available to all. This is clearly absurd in any industry, but particularly so in agriculture. One of the main justifications for public support of agricultural research and extension, in developed and developing countries alike, is the inability of farmers to search and experiment efficiently and thus to find out what technology is available.

However, knowledge of the existence of appropriate technology will not be sufficient to ensure adoption. Attitudes toward it may need to change, the hardware has to be physically available and those convinced of its value need financial resources to acquire it. One good example of this is family planning technology, where knowledge has generally outrun the capacity of the delivery systems. Similarly, local testing of the appropriateness of various items is very desirable. This in turn, will require more local agricultural research stations to accept responsibility for adaptive research and technology testing. Nevertheless, knowledge is obviously a necessary prerequisite to adoption, and publications such as this have an important part to play in information dissemination.

The impact of technology

Selection of technology for inclusion in this book does not imply endorsement of a particular product. Indeed, supporters of the appropriate technology concept often have an ambivalent attitude to new technology. New technology always changes the system and in particular it is likely to change who benefits from it. Appropriate technology advocates believe the kind of cheap, simple, small-scale, locally produced, reliable or at least mendable technology will increase incomes and improve or at least avoid worsening income distribution. This is possible, but it is still hard to prove that any technology has the ideal intrinsic qualities that will somehow create wealth and at the same time favour the poorest groups in society. On the contrary, experience shows that the income-distribution consequences of change are generally unpredictable. Since new technology normally requires access to resources, it generally favours the better off; the mode of use of technology, and thus its impact, is not a readily visible quality.

To reject all modern technology on grounds related to fears about income distribution is to argue like the elderly man who said that 'if God had meant us to fly He would not have given us railways'. Societies must accept the benefits of new technology and devise means to reduce the social costs associated with any worsening of income distribution — the greater the gain in aggregate income from innovation, the easier this should be to achieve. We are aware that the direct users of the book will seldom be the small farmer client that the contributors and compilers generally have in mind in selecting equipment. But those who have access to this book, such as extension officers, government officials,

FAO

Well-designed hand tools can reduce drudgery: harvesting in Morocco.

teachers and local leaders must give guidance with care and with wisdom.

How to choose

In selecting new technology, either for testing or promotion, numerous criteria can be devised to aid judgement. These will include the degree of technical effectiveness, financial profitability, the economic and social returns, health and safety factors, the administrative and legal compatibility with existing conditions. The criteria will not necessarily be independent or even compatible. A financially profitable piece of technology may depend upon underpriced foreign exchange or tax allowances and be economically unattractive. It may substitute capital expenditure on machinery for labour and be socially unattractive. A particular criterion such as technical efficiency, may have several elements to aid judgement — such as the technology's simplicity and labour-intensity, its ecological appropriateness, its scale and flexibility, its complementarity with existing technology and so forth. These elements are not inherently equal and in some circumstances one will be regarded as carrying most weight, in other circumstances another. Choice of technology is a matter of judgement and all the modern aids for technology assessment, for cost-benefit analysis and the like cannot hide this fact. Analysis is an aid to and not a substitute for judgement; the social consequences — which are agonising — must be weighed against the various real benefits that are apparent.

The technologies presented in this book reflect the belief that whilst all technology will alter the economic status of large numbers of people (often in the direction of greater inequality of income, greater commercialization, more wage labour and increasing landlessness) some technologies are more likely to do so than others. You will find few tractors or combine harvesters in this book, for example, but great emphasis on, for example, animal-drawn tool-bars and powered threshers. Technology varies in its degree of reach-down to the low-income farming groups who, if they are not the main target of rural development, are from our viewpoint a key component. The cost of lost output through using less efficient equipment — hand pumps rather than tube-wells, resistant seed rather than crop protection, hand tools rather than tractors, small livestock rather than cattle and buffaloes — is small. Indeed, the productivity of labour-intensive gardening and allotments can often exceed that of modern capital-intensive farming systems — as was shown in Britain during and after the Second World War. Whilst situations do occur where demand for increased food supplies force governments to chase home-produced food without too much thought about the social impact of the production system, such dire circumstances are rare. They might occur where the bulk of low-income people are food purchasers — urban dwellers and landless rural labourers, and in these cases large-scale, capital-intensive state or private farming with the most modern technology, might be justified. But it is only rarely that the trade-off between technical and economic efficiency and equity criteria is painful. Research in many countries has shown that modernized peasant-based systems are generally equally or more efficient and to most views more equitable, and thus it is the small farmers who are seen as the main beneficiaries of *Tools for Agriculture* — even if they are unlikely themselves to be the main readers of this book.

Feedback

Whilst there are a number of people who know and understand the hardware described in this book, there is less understanding of the ways in which technologies are 'delivered', or options presented to the small farmers themselves. ITDG is therefore always pleased to have critical and appreciative feedback — from the aid agencies, extension workers, credit agencies, schoolteachers, businessmen, politicians and others who use this text, on the content and format, equipment that is missing, new problems, the effectiveness of the equipment, the service of the manufacturers, and new ideas for delivery. The hardware available grows rapidly in diversity and power, but, just like computers, it will be useless without the software support. In the case of agriculture, technology software stems from the efforts of interested individuals and groups who are close to the small farmers. We look forward to hearing from you!

Ian Carruthers
Wye College, University of London

Modern technology is easily applied, if one has the resources: fertilizing maize.

BEEKEEPING

DR E. CRANE

Collecting honey from a traditional Egyptian hive.

Throughout history, and in all regions, beekeeping has been a specialized occupation of certain communities or families, remaining a mystery to the population as a whole. This is still true today, although now there are also large commercial beekeeping enterprises, and state and collective bee farms. The range of beekeeping operations in the tropics and subtropics is greater than anywhere else — from primitive honey hunting to some of the largest beekeeping enterprises in existence. Most of the honey exported onto the world market is produced in the subtropics.

Honey production involves both stock rearing (bees) and the handling and processing of food (honey). Widely differing items of equipment are therefore used, at various technological levels. In general a knowledge of beekeeping is necessary in order to understand the design and use of the equipment.

In addition to the different technological levels, beekeeping in the tropics and subtropics uses bees of different species and races, each with its own characteristics. Most beekeepers in temperate-zone countries are familiar only with the European honeybee *Apis mellifera*.

BEES KEPT IN THE TROPICS AND SUBTROPICS

Some beekeeping equipment must be precision-made according to the size of the worker bees. Bees build parallel combs at a precise distance apart, depending on the body size of the worker, and frame hives will not succeed unless they conform to this distance.

European and Mediterranean bees

The most widely used bees in the world are European *Apis mellifera*. Most of the equipment sold, and thus most of the entries in this catalogue, is for use with this bee. The modern movable-frame hive was developed in the last century for this temperate-zone bee. It was not designed for tropical honeybees, and much time and effort have been wasted in the past by trying to manage tropical bees in the same type of hive, and by the same methods.

Various types of European bees were taken overseas, and their descendants are the bees used in most parts of the New World, where there are no native honeybees. European bees are now widespread in the Americas, Australia, New Zealand, and some of the Pacific Islands. In the Mediterranean region — including Africa north of the Sahara — fairly distinct types of *Apis mellifera* are native. Some of them (in Israel, for instance) are now largely replaced by more productive bees of European ancestry. However, except in an isolated oasis or island, such replacement must be a continuing process, since new young queens are likely to mate with native drones, giving hybrid offspring of little use.

African bees

Tropical Africa also has native *Apis mellifera*. They are slightly smaller than European *Apis mellifera*, and their behaviour is notably different. They are more readily alerted to fly off the comb and to sting, and when one bee stings, others are attracted to sting at the same place. Colonies are liable to abscond from their hives if disturbed, and in some areas the colonies migrate seasonally. These are paramount factors governing bee management and hive design.

In Madagascar the native honeybee is a subspecies *Apis mellifera unicolor*, and this bee was introduced in past centuries to islands previously without honeybees, including Mauritius and Réunion. European bees are introduced successfully (and of necessity continuously) into these islands. In the very south of mainland Africa is *Apis mellifera capensis*.

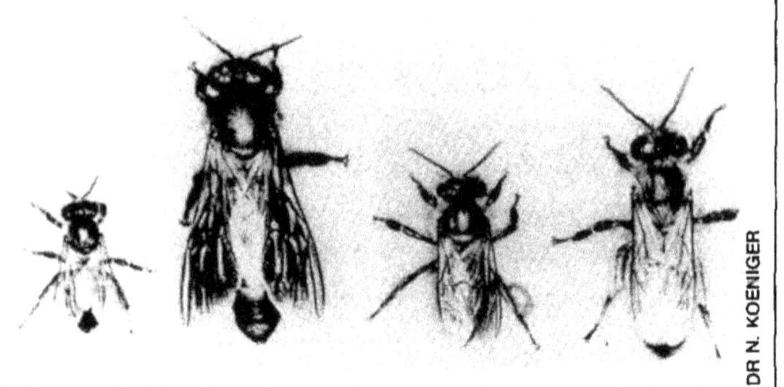

DR N. KOENIGER

Workers of the four honeybee species: from left 'Apis florea', 'Apis dorsata', 'Apis cerana', European 'Apis mellifera'.

Asian bees

Asia is the most complex continent with regard to honeybees, there being three native tropical species, *Apis cerana*, *Apis dorsata*, *Apis florea*. There are *Apis mellifera* native in the west (Turkey, the Levant, Iran, Iraq, etc.), and European *Apis mellifera* has also been introduced in many places elsewhere. *Apis cerana*, the Asiatic hive bee, looks like a smaller version of *Apis mellifera*. In India and elsewhere it is kept in small frame hives. The size of *Apis cerana* varies more than that of *Apis mellifera*. The smallest are found in parts of lowland tropical Asia, and the largest in the western Himalayas; the latter are about the size of *Apis mellifera*, and European-type hives and fittings can be used for them.

In eastern Asia *Apis cerana* has spread northwards as far as China, Korea, Japan and the Far East of U.S.S.R., i.e. into the north temperate zone. *Apis mellifera* has been introduced into these same regions, and is now used in many agricultural areas where it is much more productive than *Apis cerana*. It is the basis of the beekeeping industries of the countries concerned. Beekeeping with *Apis cerana* tends to be a separate activity, often employing traditional fixed-comb hives and management methods, confined to hill country with native flora, where *Apis mellifera* would not do as well.

Apis dorsata and *Apis florea* build a single comb in the open, and cannot be kept in enclosed hives. Both live only in the tropics of Asia. *Apis dorsata* is the largest of the honeybees; its comb may be a metre or more across, and it yields much honey. The honey is harvested by honey hunters, as described below.

Bees in Latin America

Latin America has seen a great change in beekeeping during the past thirty years. European *Apis mellifera* was used previously, but in 1956 some tropical *Apis mellifera* queens were introduced from South Africa; their offspring hybridized with the *Apis mellifera* already there, and proved dominant over them; they were *tropical* bees, whereas the European bees were not. These 'Africanized' bees have now spread throughout much of South America and well into Central America. They still have the tropical African characteristics, including high 'aggressiveness'. This has altered management practices but has also increased honey yields.

Migratory apiary in Australia, where Langstroth hives are used.

DIFFERENT LEVELS OF HONEY PRODUCTION

Honey hunting

Certain communities in Asia and Africa get much honey by hunting wild nests of honeybees in trees and rocks. In tropical Asia all of the large honey harvest from *Apis dorsata* is obtained in this way. Honey hunters reach the nests by ladders, or from a rope let down from the top of the cliff above the nests. Although honey hunting is a widespread and hazardous occupation, very little attention has been given to improving the equipment used, and none is on sale as such — so it is not recorded in this catalogue. It may include specially shaped knives to cut the combs out, and appropriate wide containers to catch the pieces of comb and carry them home. A smouldering bunch of twigs, grass, etc., is used to smoke the bees.

Collecting honey from nests of other honeybees (often in trees) is somewhat less dangerous. Combs taken from the nests are put into barrels, gourds or baskets, all locally made. *Apis florea*, whose range extends into China, and as far west as Oman, is used for a primitive form of beekeeping in Oman, but again, no equipment is on sale.

Traditional hives

Tropical Africa has a rich tradition of beekeeping in hives made locally from a log or bark, earthenware, or basketry of various types. These hives often show a high level of craftsmanship, and some communities have developed careful and ingenious methods of taking honey without killing the bees. Equipment is made locally, and there are no 'suppliers'. Log and box hives are used for *Apis cerana* in Asia. There are other fixed-comb hives, usually with no provision for bee management, and therefore needing no equipment purchased from a supplier.

Another group of bees, the stingless bees or Meliponinae, yield modest amounts of honey in tropical America, and to a smaller extent in Africa. The nests are hunted to obtain the honey, as they are also in the tropics/subtropics of Asia and Australia. Particularly in Latin America, the bees have been kept in log and pot hives using methods probably unchanged for centuries, and also in a few 'improved' hives — but these are not stocked by suppliers.

Both honey hunting and traditional beekeeping are carried out with equipment made locally from local materials, at virtually no cost except for the time taken, and following the experience of previous generations. On the other hand, most beekeeping development programmes are based on improved techniques, and on locally manufactured or purchased equipment. They can give much higher yields, but the introduction of high-cost purchases in place of home-made equipment from local materials changes the nature of the enterprise.

Traditional beekeepers may use specialized tools that they cannot themselves produce, for instance knives and other metal implements for removing combs from long cylindrical mud hives. The picture shows a set used with mud hives containing *Apis mellifera lamarckii* in Egypt. Such tools are long-lasting, and a blacksmith would be able to copy them when needed.

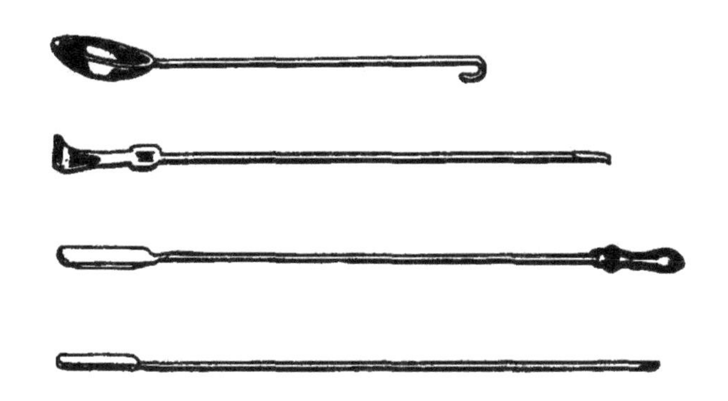

Tools used with traditional Egyptian hives.

Modern movable-frame hives

There is a great gulf between harvesting honey in the traditional ways — whether by hunting or from hives — and 'modern' beekeeping, for which equipment is purchased. Modern beekeeping is based on the movable-frame hive devised by the Rev. L.L. Langstroth in the USA in 1851. This hive was the culmination of much experimentation in Europe and North America during previous decades. It uses rectangular wooden 'frames' to support the combs the bees build. In a natural nest, combs are spaced so as to leave the same distance (a bee-space) between comb surfaces facing each other. The wooden frames are similarly distanced from each other so that combs are separated by a bee-space. They are suspended on 'runners' like files in a suspension filing cabinet. They are movable (i.e. the beekeeper can remove any one frame at will): they are also suitably distanced from the inside walls of the hive, so the bees 'respect' this distance and do not build comb across it. If a larger space is left, bees will build more comb in it; if less space, they will attach the frames to the hive walls.

Nowadays a hive is made up of several superimposed hive boxes, each with its complement of suspended frames, and frames in one box are also distanced from those above and below by a suitable bee-space. Thus each frame and hive box must be made to quite precise dimensions. Each box must fit exactly on to the one below, with no gaps through which bees could enter or leave.

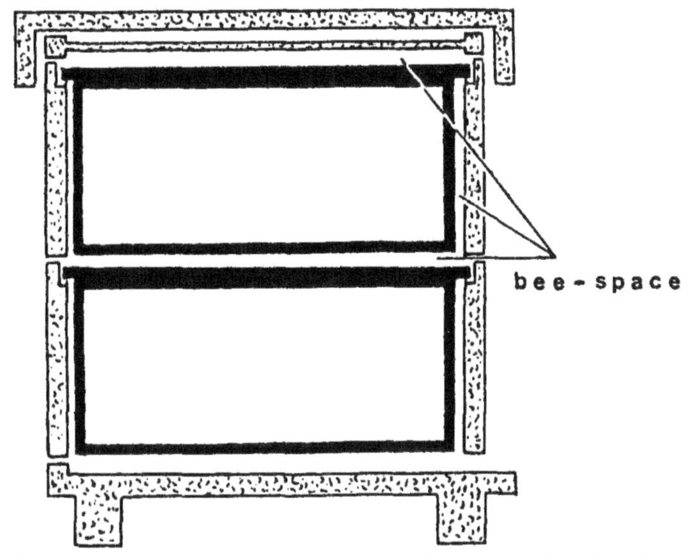

Cross-section of Langstroth movable-frame hive showing bee-space.

Intermediate movable-comb hives

In the last few decades 'movable-comb' hives have been developed, which are at an intermediate level of technology. Instead of the frames, there is a series of wooden top-bars only, suspended on runners and spaced similarly to the frames. This spacing is the only precision measurement in the hive. The sides of the hive slope inwards towards the bottom. The bees build combs downwards from the top-bars, but do not attach them to the sloping walls. These hives are made as a single,

Intermediate movable-comb hive.

extra-long box accommodating about 30 frames, instead of several movable-frame hive boxes, each holding about 10 frames, and used one above the other.

Another intermediate hive is also a long hive, but it has vertical sides, and is fitted with partial frames (top-bars with end-bars). The partial frames are of a size that will fit a Langstroth hive, so beekeepers can progress from movable-comb to movable-frame beekeeping by transferring their partial frames into Langstroth hive boxes. A few beekeeping equipment manufacturers make these intermediate hives, but otherwise almost the whole of the beekeeping equipment on sale is for movable-frame beekeeping with *Apis mellifera*. This beekeeping is, after all, the basis of the world's honey-producing industry. Movable-frame hives and fittings for *Apis cerana* are on sale in India.

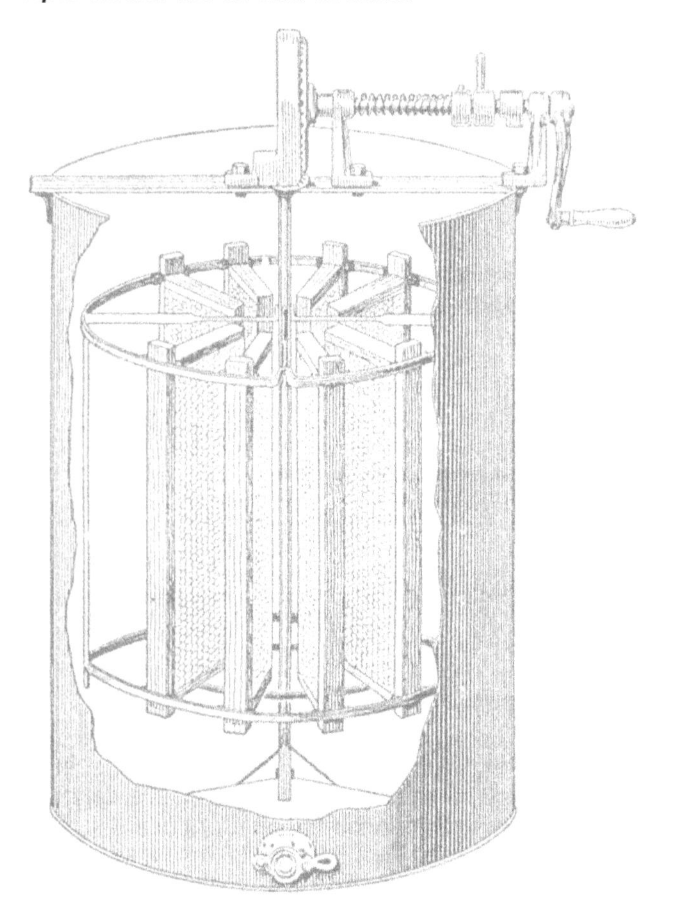

Radial extractor for spinning honey out of the cells in the comb.

PURCHASE OF EQUIPMENT

When equipment should be purchased

Beekeeping is carried out on a small scale (up to 20 hives), medium scale (20-200 hives) and large scale (200-50,000 hives under one control). When thinking in terms of equipment appropriate for different scales of operation, the position is made easier by the fact that *the hive is the unit in beekeeping*. In movable-frame beekeeping the hive box is the unit within the hive. Increasing the scale of the operation requires more hives, and more hive boxes and fittings, more bee suits and smokers, more hives set aside for queen rearing, and so on, but it does not use larger hives. Increasing the honey production per hive requires more frames and boxes for each hive.

Increasing the total honey production directly affects the scale of the honey-handling equipment needed. Whether beekeepers produce 100kg of honey a year or 100 tonnes, they must follow similar procedures in handling it, but the equipment must be appropriate for the amount handled. In general, fairly small-scale equipment is described here, on the grounds that beekeepers who have moved on to large-scale operation are likely to have more knowledge of what is available.

Many users of this catalogue are likely to live and work in areas outside those where movable-frame *Apis mellifera* beekeeping is the rule — or indeed is appropriate. For this reason items in the catalogue are arranged in the following order

- Useful for any type of beekeeping, pages 221, 224: protective clothing, smokers, hive tools;
- Used only in movable-frame beekeeping, pages 222-226: hives and fittings, etc., (movable-comb beekeeping, page 222);
- Used for handling hive products, pages 227-230: honey and beeswax extractors, etc.

Some of the equipment for handling hive products is designed for use with the larger yields obtained with movable-frame hives. But, provided a sufficient quantity of honey or wax is to be processed, much of it could be used for other types of beekeeping.

In beekeeping, benefits from using the more expensive precision-made equipment are based on the fact that such equipment allows more and better *management* of colonies of bees. The modern beekeeper aims to manage his or her colonies so that they do not swarm, and their energy is diverted instead to storing more honey which he or she can harvest.

Advantages of purchasing equipment

Purchase of the equipment listed, from reliable suppliers, has the following benefits:

- The equipment is made in large quantities, using machines that guarantee precision where this is needed
- The equipment is made of appropriate and well prepared materials (wood, metal, plastic, etc.).
- Some of the materials used are not obtainable in every country. Examples are high quality stainless steel and moulded polyurethane. For the latter, very large numbers of each article must be produced and sold to cover costs.

However, few of the specialist suppliers listed below will be in the same country as the would-be purchaser. It may therefore be necessary to buy a specific piece of equipment from a foreign country. The following are some of the circumstances which would make a purchase from a foreign country especially useful:

- The equipment is manufactured from materials superior to those available locally, for example from spring steel or stainless steel.
- The design is superior to local design, e.g. honey gates (valves for obtaining a controlled flow of honey from a honey tank, etc.).
- The operation of the equipment depends on critical factors not easily understood from a description, and hence not easily copied by a local manufacturer, e.g. some beeswax processors and pollen traps.
- Manufacture is viable only if large numbers are produced, e.g. moulded plastic honey tanks and containers.
- The precision required, for example for making frames for hives, is not available locally.
- Purchase helps to raise the standard of bee management, e.g. an effective smoker, or to raise the quality of honey and beeswax, e.g. fine-mesh honey strainers.

On the other hand, readers should be warned against purchasing unnecessary gadgets. Some beekeeping suppliers list a few such gadgets to satisfy a local demand — created possibly by publicity in the beekeeping press — but such items do not form part of the basic equipment which is needed everywhere, and their use may waste much time as well as money.

The advantages listed apply to competently run groups and enterprises (including development projects) which have access to capital, revolving funds or loans. They also apply to an individual with some capital, provided he or she has gained enough knowledge to make full use of the equipment, or can be sure of getting instruction whenever he or she needs it. For this individual, acquiring such equipment may be an opportunity to be seized, bringing considerable benefit.

It is different for poor peasant farmers who win their livelihood by using their environment to the best advantage for their crops and animals. A factory-made 'improved' hive of any sort is an alien intrusion in this environment. Unless they can receive constant support in their hive management from outside, they may revert to the familiar hives they made themselves, or they may use the new hive *as though it were* a traditional hive, and thereby forego any benefits from it.

Indicative costs and benefits

Costs vary from country to country, and according to the quality of materials and workmanship — both of which affect the precision which is essential to effective modern beekeeping. The only manufacturer to quote prices for three types of hive (movable-frame, top-bar, and long African), is John Rau & Co. Ltd. in Zimbabwe. A frame hive with brood box and two honey supers (all fitted with frames) is quoted at (Zimbabwe) $50, and a top-bar hive or a long hive at $26. A traditional hive made by the beekeeper from local materials could cost little or nothing.

A competently managed movable-frame hive might yield more or less twice as much honey as a top-bar hive or long hive, and ten times as much as a traditional hive. If the capital and the competence are assured, in many circumstances an upgrading of the level of operation to the most efficient available will give more than a proportionately higher return. If not, then beekeeping even at the traditional level provides extra food, and modest amounts of honey and wax to barter or sell, with virtually no capital outlay.

HEALTH AND SAFETY

Accidents constitute the chief hazard to both beekeepers and honey hunters. In one of the few quantitative studies made, the death certificates of 520 male beekeepers in the U.S.A. were examined, names being obtained from obituary notices. Only one cause killed a significantly higher proportion of the beekeepers than of males in the general population — accidents — which killed 32 of the 520, whereas only 19 would be expected. Many of the 32 died after a road accident, but one suffered fatal burns when smoking his bees, and another was asphyxiated when he used a plastic bag to protect his face from stings. Another cause of accidental death among beekeepers has been poisoning by cyanide when killing wasps' nests, or colonies of bees.

In traditional beekeeping in tropical Africa (where hives are sited in trees for safety), and in honey hunting everywhere, the greatest common hazard may be falling in attempting to reach and work at the bees' nest. In any community that harvests honey from wild nests, a honey hunter's rope is likely to be the strongest one the community possesses.

There is a real need for the development of methods and equipment for reducing the mortality among those who collect honey from *Apis dorsata*, as well as for obtaining cleaner and better quality honey from this bee. It might also be possible to make the honey hunter's lot easier by providing efficient smokers, and effective protective clothing, which would however be very hot.

In at least one area another hazard is responsible for most deaths. Honey collectors in the swampy Sunderbans forest at the mouth of the Ganges in India numbered from 913 to 1495 each year in the years between 1963 and 1972. Of these there were 96 casualties from carnivorous animals, an average of at least 1 per cent a year.

The layman might think that stings would be the chief hazard in beekeeping. But apart from the tiny minority of people who are allergic to bee venom, stings present little hazard to the health of beekeepers. Reactions are limited to local swelling and itching, and even these may be absent. Beekeepers normally acquire considerable immunity to stings, and 20 or even 50 stings on one occasion would not necessarily cause more than temporary inconvenience. The greatest number of stings known to have been received by a person who survived them is 2243; other survivors have received 500 or 600.

In the tiny minority of people who become allergic (hypersensitive), general bodily reactions occur: rash, much swelling, difficulty in breathing, and even unconsciousness. Anyone who suffers a general reaction should give up beekeeping and avoid future situations where he or she might be stung. Medical advice should be sought, and in countries where a desensitization course is available, this should be discussed with a medical specialist.

It is always better to avoid being stung, and protection against stings — especially in the eyes or mouth — is strongly recommended. Protective clothing is the first item of equipment described below.

SOCIO-ECONOMIC IMPACT OF CHANGING THE TECHNOLOGY

Beekeeping development does not need high investment or complicated technology. Simple hives can be made from a variety of natural products which are familiar to the rural populations in different parts of the world. Some are already used for traditional hives. Colonies of bees to populate new hives can be obtained by collecting swarms, or by dividing existing colonies. In some places a subsistence farmer can get a higher income from beekeeping than from all the other work he does during the year. Also, in rural areas with subsistence agriculture, beekeeping raises the social standing of successful beekeepers and, by producing honey, beekeepers broaden the food basis of the population.

Whether it is done to produce food for the family or to provide a cash crop, beekeeping allows great flexibility in the amount of time it occupies. According to the number of hives kept, it can be spare-time, part-time, or full-time. Through the formation of co-operatives, beekeeping can stimulate professional and social contacts for the benefit of an entire group of people.

In its simplest form, beekeeping needs no imported technology or investment. If the technological capability is available, the beekeepers' requirements for hives honey containers and other equipment can stimulate production by local craftsmen. At higher levels of operation, it may be necessary to import technical equipment for beekeeping, and for processing honey and beeswax.

Where beekeeping becomes a large-scale operation carried out at a modern technological level — with movable-frame hives — it ceases to be a means whereby subsistence farmers can improve their lot through use of local materials and traditional crafts. Capital investment is needed, and labour requirements will probably be minimized in order to increase profits. Honey is produced for sale in the larger towns, or for export (earning hard currency), but the lowest income groups are unlikely to benefit from this.

There is one gain from any increased beekeeping which can benefit the whole rural population. Through pollination, the food-gathering activities of bees improve both the quantity and quality of many cultivated crops. The intensification of agricultural production frequently includes a greatly increased use of fertilizers and

Table 1. The world honey industry, as represented by figures for 13 countries.

Country	1 *Colonies × 1000*	2 *Yield per colony*	3 *Total honey × 1000*	4 *Net exports × 1000*		5 *Honey per capita*	6 *Sugar per capita*
Europe						0.4	45
France	1200	12.7	18.5		– 6.7		
German F.R.	1118	12.6	15.0		– 62.9		
U.K.	212	6.3	1.2		– 20.8		
North America						0.7	49
Canada	657	51.3	34.8	+ 9.5			
U.S.A.	4275	22.8	93.0		– 37.9		
Australia + New Zealand						0.5	57
Australia	405	56.0	21.5	+ 1.1			
New Zealand	191*	30.0*	7.6	+ 2.0			
Latin America						0.1	42
Argentina	1300	25.5	28.0	+ 29.9			
Brazil	1800	13.3	22.0	+ 0.6			
Mexico	2300	25.5	64.0	+ 40.0			
Africa						0.26	11
No single country of world importance							
Asia						0.0004	7
China	5700	19.6	100.00	+ 58.1			
Japan	299	21.4	6.5		– 28.1		
U.S.S.R.	8000	23.0	190.0	+ 16.0		0.5	45
Total	27457		602.1	157.2	156.4		
World total			896.3	214.3	224.7		
% of world represented by the 13 countries			67%	73%	70%		

* from the same source as column 5

Column 1 — *Colonies* × 1000 gives the number of occupied hives in thousands in 1983.
Column 2 — *Yield/colony* gives the average honey yield in kg per colony, 1979-83.
Column 3 — *Total honey × 1000* gives the estimated total honey production for the country in 1983, in 1000 tonnes.
Column 4 — *Net exports × 1000* gives the country's estimated honey exports *less* honey imports, in 1000 tonnes, for 1982. Figures prefixed by + are net exports, and figures prefaced by – are net imports.
Column 5 — *Honey per capita* gives the estimated average honey consumption in kg per capita for the continent as a whole, from sources quoted in E. Crane, *Honey: a comprehensive survey*, published in 1975, but relating to various earlier years. Figures for Africa and Asia are less reliable than others.
Column 6 — *Sugar per capita* gives the average sugar consumption in kg per capita for the continent as a whole, from the United Nations Statistical Yearbook (1970); most figures relate to 1969.

pesticides. The latter often kill the population of wild insects that serve as pollinators of cultivated crops. The only remedy is to provide a pollination service, by moving hives of honeybees to the agricultural production areas during the flowering of the crop, and not killing the bees with insecticide while they are there.

SUMMARY OF THE WORLD HONEY INDUSTRY TODAY

Table 1 gives figures for 13 countries. Columns 1-4 are taken from the statistics of the United States Department of Agriculture (*USDA Foreign Agriculture Circular* FS3-83). The totals at the foot of Table 1 show that the data for the 13 countries represent two-thirds to three-quarters of those for the world as a whole, and therefore help to present a world picture.

Table 1 shows high honey yields per colony in Canada and Australia and low ones in Europe. It also shows the high total honey production of the large countries, U.S.S.R., China and U.S.A. (190, 100, 93 thousand tonnes, respectively). The high honey-exporting countries are China, Mexico and Argentina (58, 40, 30 thousand tonnes), and the high net importers are the German Federal Republic, U.S.A., Japan and U.K. (63, 38, 28, 21 thousand tonnes). Until 1981 Japanese imports exceeded those of the U.S.A.

The three largest exporters are thus in the subtropics, and countries in which the European honeybee is not native. All of the four largest importers are comparatively rich countries, and all are in the north temperate zone. Germany and the U.K. belong to the traditional 'bees-and-honey' region in Europe, and the U.S.A. was peopled from this region. Japan, alone, has developed as a honey-eating country since the Second World War. In the final two columns in Table 1, figures for honey and sugar consumption per capita for the continents as a whole are lower for Asia than for any other continent. This situation may change as honey production increases, but only when incomes also rise: Table 1 suggests that purchased honey is now a food of affluent societies.

HOW TO PURCHASE BEEKEEPING EQUIPMENT

Beekeepers can much more easily purchase equipment from a supplier in their own country, if it is available, than from abroad. Beekeepers are urged to try to see a supplier's equipment — if possible in use — and to discuss it directly with the supplier before any purchase is made. In different areas, paramount qualities may vary — for example suitability of hives for hot, dry conditions, maintenance of metal equipment in year-round high humidity, or resistance to termite damage. The following pages are a descriptive, illustrated catalogue of 66 types of equipment. With each description is the name and address of a specialist supplier (if possible one known to manufacture it), or an indication that it can be obtained from most general suppliers.

Some of the general suppliers worldwide are listed below, and further suppliers and manufacturers can be found in the catalogue.

Dr Eva Crane
International Bee Research Association

ARGENTINA

MIGUEL A BREJOV
Nazca 4058/74 (1419)
Buenos Aires
ARGENTINA

EL PANAL
S.A.C.I.F.I.Y.A.
Humahuaca 4229
1192 Buenos Aires
ARGENTINA

MECANIZACIÓN APÍCOLA SRL
Calle 35, No.407
La Plata (B.A.)
ARGENTINA

TERZA HNOS
S.A.C.I.F.I.Y.A
Floor 5, Corrientes 1312
1043 Buenos Aires
ARGENTINA

AUSTRALIA

JOHN L. GUILFOYLE (SALES) PTY. LTD.
772 Boundary Road, Darra
Brisbane P.O. Box 18
Queensland 4076
AUSTRALIA

PENDER BROS. PTY. LTD.
Elgin Street, P.O. Box 20
Maitland, NSW 23200
AUSTRALIA

AUSTRIA

STEFAN PUFF GmbH
Neuholdaugasse, 8011 Graz
AUSTRIA

BELGIUM

RAYMOND DE BIE
Mechelsbroekstraat 21
2800 Mechelen
BELGIUM

BRAZIL

CAPEL
Parque de Exposição de Animals
DPA
Av. Caxangá, 2200
CEP 50 000 Recife (PE)
BRAZIL

PRO-APIS LTD (PROJECTO DE APICULTURA)
Rua Almirante Larnego
38- Edifício Panorama
Caixa Postal 880, Fiorianópolis (SC)
BRAZIL

CANADA

BEEMAID
625 Roseberry Street
Winnipeg, Manitoba, R3H 0T4
CANADA

F.W. JONES & SON LTD
44 Dutch Street
Bedford, Quebec, JOJ 1AO
CANADA

CHILE

CRATE
Casilla 6122, Correo 22
Santiago
CHILE

COLOMBIA

PROAPÍCOLAS LTDA
Near Pitalito, Huila
COLOMBIA

DENMARK

ØSTJYDSK, BIAVLSCENTER ApS
Vejle Landevej 147 (A 18)
Pjedsted 7000, Fredericia
DENMARK

ANNE MARIE & BERNHARD SWIENTY
Skovbrinken 12
6400 Sønderborg
DENMARK

EGYPT

HASSAN ALLAM
17 Boutrors Street, Tanta
EGYPT

HOUSE OF BEES AND AGRICULTURAL ACTIVITIES
6 Sekket El Manah Street
Opera Square, Cairo
EGYPT

MOHAMED EZZ
7 Army Street, Cairo, EGYPT

FRANCE

APICULTEUR ALPHANDERY
Château de Brignan
84140 Montfavet (Vaucluse)
FRANCE

APICULTURE NEVIERE s.a.r.l.
BP 15
Route de Manosque, 04210 Valensole
FRANCE

CAURETTE
139 Rue La Fayette
75010 Paris
FRANCE

EUROPRUCHE
Boulevard De L'Industrie
Z.I. des Loges
53940 St. Berthevin-les-Laval
FRANCE

LEROUGE
91 Rue Mangin
60130 St. Just-Chaussée
FRANCE

MAX MENTHON
36-38 rue du Commerce
74200 Thonon-les-Baus
FRANCE

CHRISTIAN NICOT
Maisod, 39260 Moirans-en-Montaigne
FRANCE

ETS THOMAS FILS SA
65 Rue Abbé Georges Thomas
BP No.2, 45450 Fay-aux-Loges
FRANCE

GHANA

TECHNOLOGY CONSULTANCY CENTRE
University of Science and Technology
University Post Office, Kumasi
GHANA

GREECE

MELISSOKOMIKI
57 Makrygianni Street
Nea Chalkidon, Athens
GREECE

HUNGARY

HUNGARONEKTÁR ORSZAGOS
1054 Budapest, Garabaldi u.2
HUNGARY

INDIA

ALL INDIA BEEKEEPERS ASSOCIATION
1325 Sadashiv Peth, Pune 411030
INDIA

EASTERN SCIENTIFIC COMPANY
New B.D. High School
Ambala, Cantt 133001
INDIA

KHADI & VILLAGE INDUSTRIES COMMISSION
Carpentry and Blacksmithy Workshop
Post: Dahanlu, Dist. Thane
Maharashtra
INDIA

LOTLIKAR AND SONS
A-1/4 Pioneer Co-op Society
Panvel 410206, Kulaba M.S.
INDIA

PARAGANA BEEKEEPERS CO-OPERATIVE SOCIETY LTD.
Post Baraipur, West Bengal
INDIA

RAJ CARPENTRY WORKS
Pathankot, Dist. Gurdaspur
Punjab
INDIA

RAWAT APIARIES (Himalayas)
Ranikhet, Dist. Almore, UP
INDIA

SARVODAYA SAMITI
Gandhinagar, Koraput 764020
Orissa
INDIA

TRIPURA STATE KHADI AND VILLAGE INDUSTRIES BOARD
Post Agartala 799001, Tripura
INDIA

IRELAND

IRISH AGRICULTURAL WHOLESALE SOCIETY LTD.
151-156 Thomas Street, Dublin 8
IRELAND

MIL AN tSULÁIN
Cúil-Aodha, Magcromtha
Co. Chorcaighe
IRELAND

ITALY

LEGA SDF
Via de Cresceni 18, 48018 Faenza
ITALY

SAF, s.n.c.
Via Liguria 17, 36015 Schio (VI)
ITALY

JAPAN

AKITAYA HONTEN CO. LTD.
Kano-fuji-machi, Gifu 500
JAPAN

FURUZAWA BEE KEEPING MANUFACTURER
752 Gifu
JAPAN

GIFU YOHO CO. LTD.
Kano-sakurada-cho 1
Gifu-Shi, Gifu 500-91
JAPAN

NONOGAKI APIARY
Oku-machi
Ichinoabuyo-shi, 490-02 Maya
JAPAN

KENYA

MINISTRY OF AGRICULTURE & LIVESTOCK DEVELOPMENT
Beekeeping Branch
P.O. Box 68228 Nairobi
KENYA

MEXICO

MIEL CARLOTA, S.A.
Ap. Postal 161-D
Queretaro III, Cuernavaca, Mor.
MEXICO

MOROCCO

AGRICOLA
34 Rue Beni Amar, Casablanca
MOROCCO

NETHERLANDS

BIJENHUIS
Grintweg 273
6704 AP Wageningen
NETHERLANDS

HONINGZEMERIJ HET ZUIDEN BV
Ladonksemeg 9, Postbus 2
5280 AA Boxtel
NETHERLANDS

H.T. VAN DAM & ZN
P.W. Janssenweg 35-37
8411 XR Jubbega, Friesland
NETHERLANDS

NEW ZEALAND

A. ECROYD & SON LTD.
P.O. Box 5056
25 Sawyers Arms Road
Papanni, Christchurch 5.
NEW ZEALAND

NORWAY

HONNINGCENTRALEN A/L
Østensjøv 19, Oslo 6
NORWAY

PHILIPPINES

IMELDA'S BEEKEEPER SUPPLIES
1910 F. Tirona Benitez Street
Malate, Manila
PHILIPPINES

SPAIN

APICENTER S.A.
Vizcaya 383, Barcelona (27)
SPAIN

VICENTE MENDIPOZO
Avda España 4, Logroño
SPAIN

MIELSO S.A.
Polígono Industrial "El Mijares"
Calle No.7, Apartado 38
Almazora, Castellón
SPAIN

MODERNA APICULTURA SA
La Apartado 9.008, Madrid 28
SPAIN

VICENTE MENDI POZO
Avda Espana 4, Logrono
SPAIN

AUGUST PERPINYA
Carretera L'Hospitalet 45
Cornellá, Barcelona
SPAIN

SWEDEN

OSCAR GUSTAFSSON & CO
Biredskapsfabrik AB
4385 Tofta, 432 00 Varberg
SWEDEN

HEBE STÅL, AB
Fack 32, 684/01 Munkfors
SWEDEN

SWITZERLAND

BIENEN-MEIER
5444 Künten (AG)
SWITZERLAND

U.K.

ROBERT LEE (BEE SUPPLIES) LTD.
Beehive Works
High Street, Cowley
Uxbridge, Middlesex UB8 2BB
U.K.

R. STEELE AND BRODIE
Stevens Drove, Houghton,
Stockbridge, Hants SO20 6LP
U.K.

E.H. THORNE (BEEHIVES) LTD.
Beehive Works
Wragby, Lincoln LN3 5LA
U.K.

U.S.A.

COWEN ENTERPRISES
P.O. Box 396, Parowan, UT 84761
U.S.A.

DADANT & SONS, INC.
Hamilton, IL 62341
U.S.A.

WALTER T. KELLY CO.
Clarkson, KY 42726
U.S.A.

A.I. ROOT COMPANY
P.O. Box 706
623 W. Liberty Street
Medlina, OH 44258
U.S.A.

SUNSTREAM
P.O. Box 225
Eighty four, PA 15330
U.S.A.

W. GERMANY

CHR. GRAZE, KG.
Strümpfelbacherstraße 21
7056 Weinstadt 2, (Endersbach)
W. GERMANY

C. KOCH
Hauptstraße 67
7603 Oppenau/Schwarzwald
W. GERMANY

MÜNGERSDORFF
An St. Agatha 37, 5000 Köln 1
W. GERMANY

ERHARD & MARKUS SCHEHLE
8999 Maierhöfen/Allgäu
W. GERMANY

FRIEDRICH WIENOLD
Dirlammer Straße 20
Postfach 15
6240 Lauterbach/Hessen
W. GERMANY

ZIMBABWE

JOHN RAU & COMPANY (PVT.) LTD
2 Moffat Street
P.O. Box 2893, Harare
ZIMBABWE

PROTECTIVE CLOTHING

Every beekeeper should have adequate protective clothing, even if he or she sometimes chooses not to wear it all. The most important part to protect is the face, especially the eyes and mouth. Whether arms and hands are covered is a choice to be made by the beekeeper according to the occasion and the work to be done, and the character of the bees to be dealt with. Individual items of clothing *must* be impermeable to bee stings, and every joint between them *must* be bee-tight — if not, it could be safer to strip completely than to risk getting bees caught inside the clothing.

Modern fastening devices such as zip fasteners and Velcro have made it possible for a beekeeper to be completely enveloped in a single garment. Alternatively, separate parts may be used: veil supported by hat or hood; gloves; an appropriate coverall or boiler suit and boots, or cooler body clothes — which, however, will not give as much protection.

Except for the vizor of the veil, which must be black to give good vision, all cloth for garments worn when working with bees should be light in colour and of smooth, close-mesh material. For working with tropical African and Africanized bees, it may be best to use a veil with the outside of the wire-mesh vizor painted white, otherwise bees are likely to fly against the black mesh and obscure vision. With these bees, also, stout plastic gloves may be necessary, although they are hot and clumsy to wear. All general beekeeping suppliers stock protective clothing, but it is worth seeing *and trying on* different types, to find out what is suitable for you and for the conditions under which you work. The outfit shown is one used for working with 'aggressive' tropical African bees.

If, in spite of precautions, you find you have a bee inside your protective clothing, go well away from the bees before you investigate. A similar rule applies to removing the clothing.

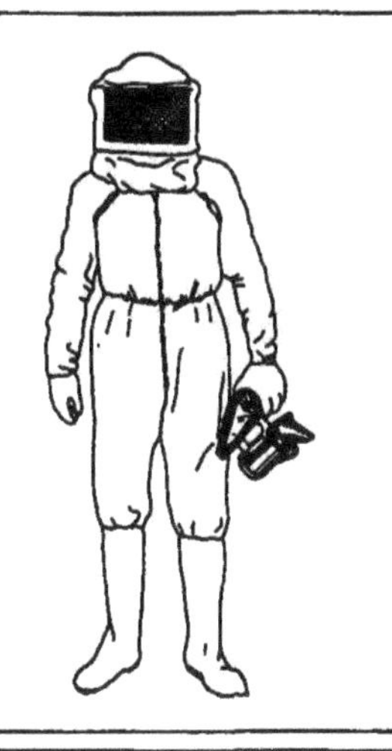

GLOVES AND GAUNTLETS

Gloves (upper illustration) should be light in colour, soft, and sufficiently well fitting to allow the wearer to work delicately when moving frames, etc., in order not to disturb the bees. The material covering the hands should be impervious to stings, and soft leather is ideal; the wider gauntlet part can be of close cotton weave. The upper hem of the gauntlet is elasticated, to be worn over a long sleeve. In no circumstances wear black gloves. Rubber gloves are sometimes advertised, but they are hot and can be clumsy. On the other hand thin cotton gloves are easily penetrated by a bee's sting.

Some beekeepers prefer to wear gauntlets only (lower illustration), in which case the lower hem is also elasticated and fits snugly over the wrist. Either gloves or gauntlets may reach below or above the elbow as required.

Available from:
GENERAL SUPPLIERS

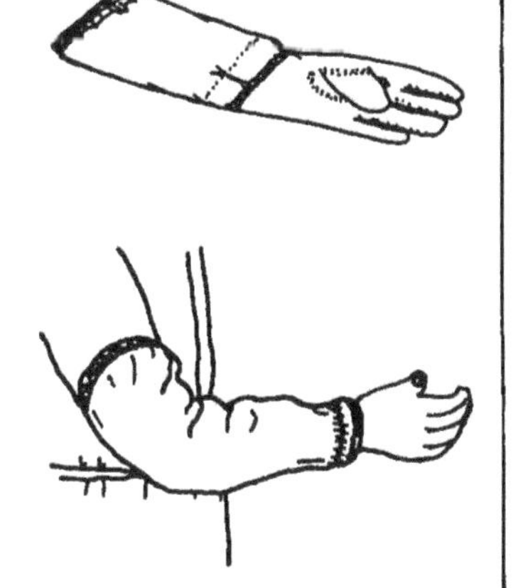

BOOTS

Many beekeepers tuck trouser bottoms into gumboots. They can be purchased at a shoe shop. Alternatively, trouser bottoms are tucked into smooth, light-coloured socks worn with shoes.

You can be stung badly round the ankle through lack of care in ensuring a bee-tight join in the protection there, and bees inside a dark space instinctively run upwards.

Available from:
GENERAL SUPPLIERS

HAT AND VEIL

The choice must depend on the type of work to be done, the temperature and wind, and personal preference. The drawing shows a folding veil in which the vizor is made up of 3 rigid sections of black wire mesh. The tapes at the front are tied round the waist in such a way that the bottom edge of the cloth is drawn tightly against the clothes beneath; alternatively the cloth below the veil can be tucked inside a sleeved jacket at the neck. The brimmed hat shown is soft, but a rigid brimmed hat (with ventilation slits if wanted) is preferred by many. The veil may be integral with a cloth hat, or separate, and held over the brim by an elasticated hem at the top.

Woven horsehair or nylon net is used for the vizor in light-weight veils. This is satisfactory, except that in windy weather it may blow against the face or neck.

Available from:
GENERAL SUPPLIERS

COVERALL

A standard coverall can be used, of a white close-weave material. Custom-made bee suits incorporate elasticated wrists and trouser cuffs. One maker sells coveralls (illustrated) of rip-stop nylon for working with Africanized bees that sting readily. They are large enough to be worn over clothing and are thin; they are reported to be 'bee-secure' although hot.

These are made by:
Mrs D OLSEN
115 South First East
Providence, UT 84332
U.S.A.

Coveralls (and other such clothing for bee work) should be washable, and washed as often as necessary. This is not only to remove any gross dirt, but to remove odours to which bees might respond by stinging, and to minimize the possibility of carrying disease infection from one apiary to another.

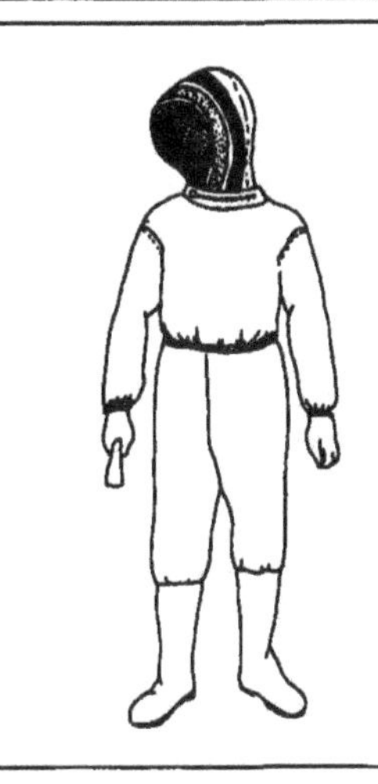

ALL-IN-ONE SUIT WITH HOOD

The drawing shows a two-piece suit, but it can be purchased as a single coverall. The wrists are elasticated. The hood is attached by a zip, (and sewn on at the back of the neck), and can thus be thrown back when not required, without removing the suit. The vizor is of black nylon net, and is kept off the face by nylon boning round the edges and by the self-supporting hood. Garments are made by:

B.J. SHERRIFF
Five Pines, Mylor Downs
Falmouth, Cornwall TR11 5UN
U.K.

By tradition, hoods have been used in certain countries such as the Netherlands, and there has recently been a swing towards them in some other countries. If possible try on a veil with a hood and with a hat, to see which you prefer.

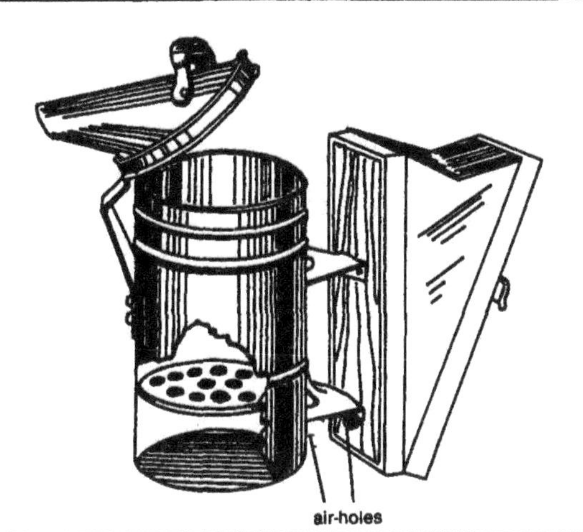

SMOKER

A good smoker is essential in beekeeping with frame hives or top-bar hives. In traditional beekeeping, smouldering twigs or grass are used to smoke bees, but this does not give the directional flow of cool smoke that is most effective, and best for keeping the bees quiet. (The bees respond to the smoke by gorging themselves with honey, and are then less likely to sting.) Some traditional beekeepers and honey hunters would probably find a modern smoker very helpful.

The metal fire box on the left has a directional funnel hinged to the top, which allows the fuel to be inserted. The fuel is kept off the base of the fire box by a perforated metal shelf above an airhole. The bellows on the right, which contain a spring, are used to blow air into the fire box through two holes opposite each other.

The aim is to produce a large and steady supply of cool smoke from the funnel without the need for frequent refuelling. Success depends on the design of the smoker and the use of a large fire box (say 25cm high and 12cm diameter), and on the fuel used. According to what is available, beekeepers use old sacking, decayed wood, wood shavings or other vegetable matter, and corrugated cardboard.

It is important that only smoke, and no flame, should emerge from the smoker, and that the fire should be extinguished immediately after use.

A few suppliers offer a smoker with a clockwork mechanism to maintain a constant flow of smoke, but such a device is not necessary.

Available from:
GENERAL SUPPLIERS

MOVABLE-FRAME HIVES

MOVABLE-FRAME HIVES FOR 'APIS MELLIFERA'

Types of movable-frame hives that are in wide enough use to be considered appropriate are Langstroth, which is the most widely used throughout the world especially in English speaking countries, and Dadant or Dadant-Blatt. Both these are for the European bee *Apis mellifera*.

In both, the bee-space between hive boxes is at the *top* of each box. This is preferable to a bee-space at the bottom, in which case frames are flush with the hive box at the top. With a top bee-space, a flat wooden cover, (e.g. to support a feeder), can be placed directly on the top hive box. (With a bottom bee-space a cover must have a frame below it to lift it above the top of the frames.) Also with a top bee-space, one hive box can be slid into position on top of the one below without crushing the bees.

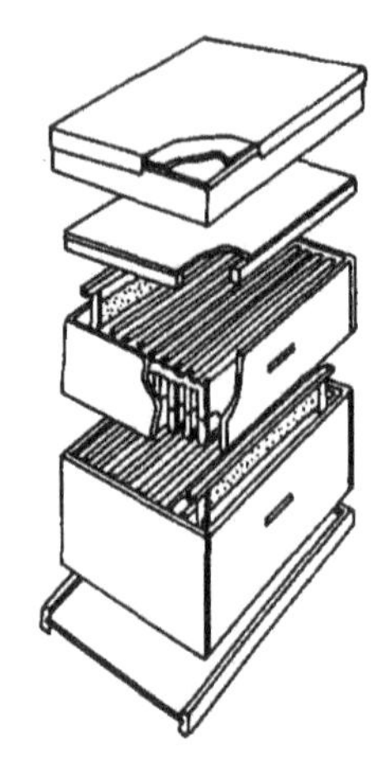

LANGSTROTH HIVE

This is the most widely-used hive in the world. The frames are separated from the hive wall (and from each other) by a bee-space.

The illustration on the left is an exploded view of the Langstroth hive, showing the parts in detail (from the bottom): bottom board, brood box or chamber, super or honey chamber, inner cover, roof. Most Langstroth hives have boxes to accommodate 10 frames, but 8-frame and 12-frame hives are also made.

Standard dimensions, and certain details of design, vary slightly from country to country, and it is therefore wise to purchase all hives and hive fittings from one supplier. Langstroth hives are sold by general beekeeping suppliers. In addition, one firm in Egypt specialises in their manufacture, and two general suppliers in India produce them.

MAKHTAR HAMED YASEEN
1 Aziz fahmy Tanta
Garbeya Governorate
EGYPT

ALL INDIA BEEKEEPERS ASSOCIATION
1325 Sadashiv Peth
Pune 411030
INDIA

RAWAT APIARIES
Ranikhet
Dist. Almore
U.P.
INDIA

MODIFIED DADANT AND DADANT-BLATT HIVE

This hive is on a similar principle to the Langstroth, but has eleven deeper frames. They are used very successfully by some large-scale beekeepers. The greater weight of each box when full makes them less generally popular, and the extra size is of no advantage unless bees can be managed appropriately. C.P Dadant, the originator of this hive, was born in France and wrote in the beekeeping press of France and other countries. As a result, a variant of this hive (sometimes known as Dadant-Blatt) is used in many French-speaking countries.

ETS THOMAS FILS SA
65 Rue Abbé G. Thomas
BP No. 2
45450 Fay-aux-Loges
FRANCE

LEGA SDF
Via Armandi 19, 48018 Faenza
ITALY

INDIAN STANDARD HIVE FOR 'APIS CERANA'

Hives on the same principle as the Langstroth and the Dadant-Blatt are manufactured for use with the smaller Asiatic hive bee *Apis cerana*; each hive box usually accommodates 9 frames. Beekeeping suppliers in India manufacture (or supply) these hives.

SEE INDIA GENERAL SUPPLIERS

TOP-BAR HIVES

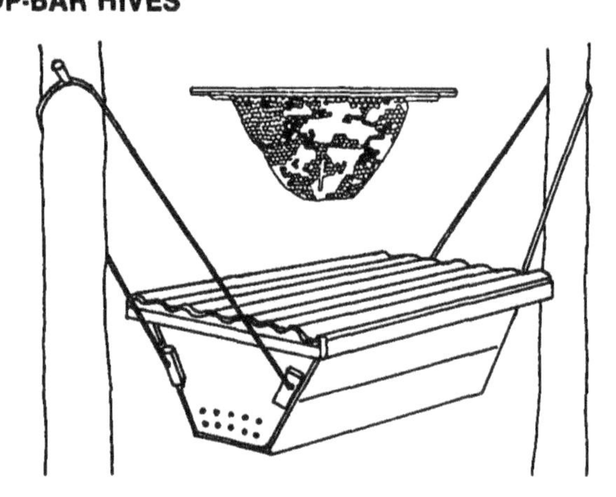

Top-bar hives are 'movable-comb' hives; they have no frames, but properly distanced top-bars. The bees build combs down from the top-bars, but they do not attach them to the hive walls, which slope inwards towards the bottom.

KENYA TOP-BAR HIVE, FOR TROPICAL AFRICAN 'APIS MELLIFERA'

This design was developed in Kenya before and during the Canadian International Development Agency project (1971-1982). Internal measurements are 88.9 × 44.3cm at the top and 88.9 × 18.9cm at the bottom, height 28.6cm. It has a complement of 28 top bars 3.2cm wide and 48.3cm long, supported by runners. Top-bars touch each other and there is no space between them. This is an important feature when handling tropical African bees, since only one bar-width is open at once, and this can be continuously smoked, so that flight by the bees (and stinging by them) is minimized.

The drawing (above) shows the entrance holes, roof, and suspension method of support — to prevent damage by ants and other enemies.

These hives are manufactured by:

MINISTRY OF AGRICULTURE & LIVESTOCK DEVELOPMENT
Beekeeping Section
P.O. Box 274, 68228 Nairobi
KENYA

They are also sold by:

BROTHER BURKE
Farmer Training Centre, Mola
KENYA

JOHN RAU & COMPANY (PVT.) LTD.
2 Moffat Street
P.O. Box 2893, Harare
ZIMBABWE

The following firm will make top-bar hives to order:
BUDGET BEEKEEPING
Gillbrow Apiaries
Kirkandrews-on-Eden
Carlisle CA5 6DW
U.K.

MODIFIED LONG HIVE

This hive was developed from the Kenya top-bar hive, and has been used there and in Tanzania. The sides are vertical, and each top-bar has 2 end-bars, but instead of a bottom-bar like a frame, a horizontal strut is fixed between the two end-bars; with them, it gives support to the comb. The Kenya top-bars fit this hive, and the partial frames can be used in a standard Langstroth hive, so the 'long hive' provides a useful step in advancing from the top-bar hive to a frame hive.

The hive (below) is supplied by:

For tropical African bees:
JOHN RAU & COMPANY (PVT.) LTD
2 Moffat Street
P.O. Box 2893, Harare
ZIMBABWE

For European bees:
AMERICAN-KENYA RESEARCH AND DEVELOPMENT CORPORATION
1204-2956 Hathaway Road
Richmond VA 23225
U.S.A.

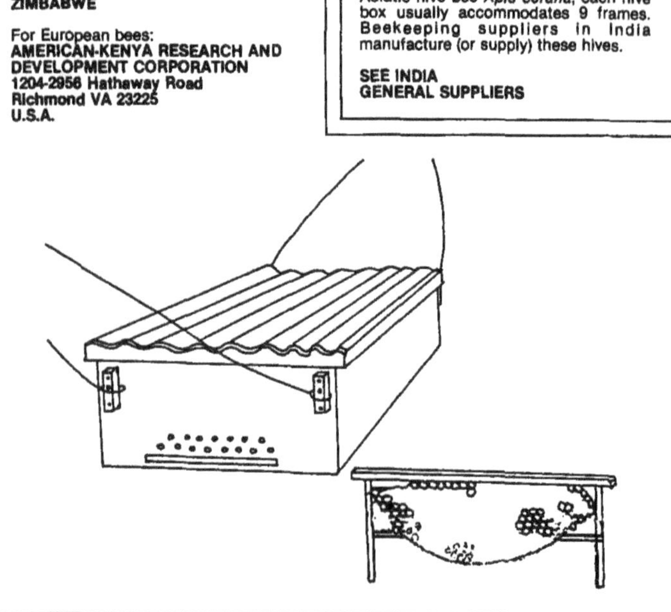

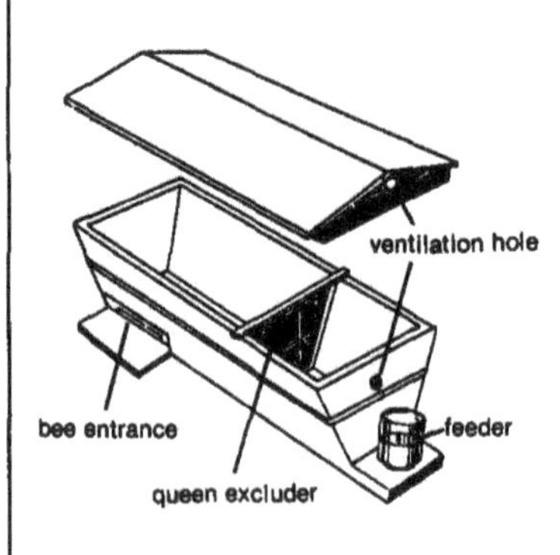

TOP-BAR HIVE FOR 'APIS CERANA'

Several attempts have been made to use top-bar hives for *Apis cerana* in Asia. The hive described here (illustrated left) is two-thirds (linear) the size of the Kenya hive above. It was designed by the late Father B.R. Saubolle, Kathmandu, and is currently being distributed in Nepal under a UNICEF/Agricultural Development Bank programme. It is suspended, for the same reasons as in Africa. It has full-width top-bars, although *Apis cerana* is very little inclined to sting. The slit entrance is taken from an earlier type of the Kenya hive, which was discarded there in favour of a series of holes (as in the Kenya hive above) which the bees can more easily protect.

A strong wire queen excluder is provided with this hive, which is made by:

GANA FURNITURE
Gana Bahal, Kathmandu
NEPAL

OBSERVATION HIVE

Many beekeeping supply firms manufacture a tall narrow observation hive, in which 2 or 3 frames are mounted one above the other, so that both sides are visible through the glass. These hives are excellent educational aids, but it can be difficult to keep the bees in good condition, especially in hot weather.

The drawing on the right shows a simpler hive in which bees build their own comb from a small piece of foundation (top left of box). The manufacturer below provides detailed drawings and instructions for assembling the hive from the kit supplied.

The hive can be populated with bees from a special travelling box, through the flexible tube (bottom left). The queen is introduced in the queen cage (top left). But do not order live bees except from within your own country.

HERMAN KOLB
P.O Box 183, 737 West Main
Edmond, OK 73034
U.S.A.

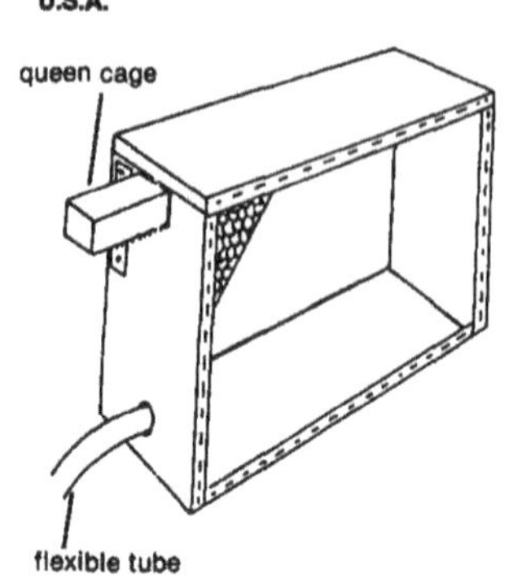

HIVE FITTINGS

The items featured on this page are for frame hives, and all of them must be of the correct *dimensions* for the hives in which they are to be used.

FRAMES

These support the wax foundation (see p.255) and the comb the bees build from it, and maintain the bee-space gap between frames/combs and hive walls; see Frame spacers. Frames are usually made of fine-grained wood, with tongue and groove or other very strong joints between bottom and end bars, and where the end bars join the top bar. This is necessary because of the weight of honey in full combs, and the strains to which frames are subjected in bee management and in honey extraction. Available from:

GENERAL SUPPLIERS

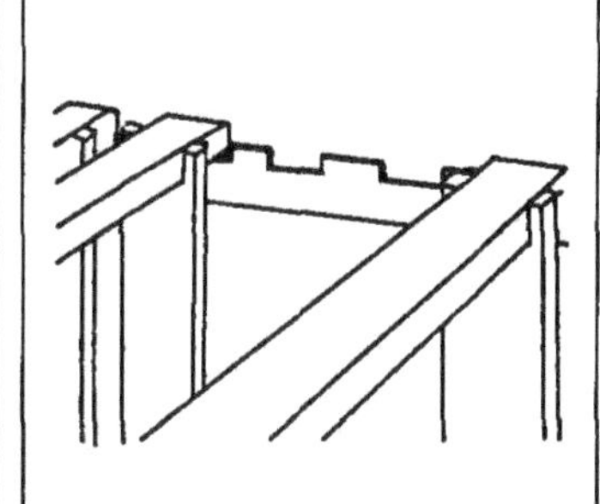

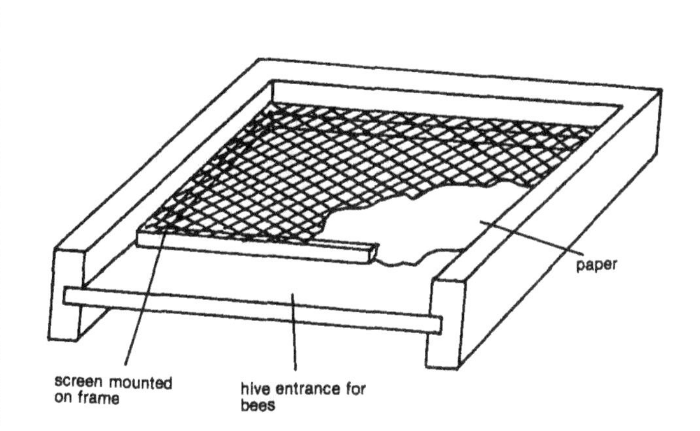

VARROA DETECTOR

This item is included in view of the publicity given to the spread of the mite *Varroa jacobsoni* to different regions of the world as a parasite of *Apis mellifera*. The drawing shows a device to be incorporated with the floorboard of any frame hive. A plastic grid is mounted above a sheet of white paper laid on the floorboard; dead *Varroa* mites fall on to the white paper and can be seen with the naked eye when the paper is inspected after a dearth period during which brood rearing is minimal.

The plastic grid (to be mounted in a frame that fits inside the hive used, as shown in the drawing) is obtainable from:

S.A.M.A.P.
1 rue du Moulin BP 1
Andolsheim
Neuf-Brisach, 68600
FRANCE

FRAME SPACERS

Some frames are spaced by their end-bars, which are widened out so that when they are touching, the combs are at the exact spacing required. Hoffman is one type. Alternatives are to put a plastic or metal 'end' on each end of the top-bar to space them correctly, or to use 'castellated' metal runners, one type of which is shown above; the frame top-bars fit into the depressions. Bees tolerate a greater variation in cell depth, and in comb spacing, in honey supers than in the brood nest. Castellated spacers are made by several firms, including the two below. The first sells many types, so send full details of what you want.

STOLLER HONEY FARMS, INC.
Latty, OH 45855
U.S.A.

B.J. ENGINEERING
Swallow Ridge, Hatfield
Norton, Worcester WR5 2PZ
U.K.

QUEEN EXCLUDER

This is a flat perforated screen, of the same size as the cross-section of the frame hive in which it is to be used. It is inserted above the brood box to separate it from the honey super above, and the slots in it are of such a size that workers can pass through but not the queen. The honey supers are thus kept free from brood.

The dimensions of the slots are critical, and vary according to the type of bee used. For tropical *Apis mellifera* they are smaller than for European *Apis mellifera*, and for *Apis cerana* they are smaller still.

Queen excluders are made in two types: (left) a flat sheet of metal or plastic with slots stamped out by machine; (right) a series of parallel wires soldered to cross-strips, and the whole mounted in a wooden frame. The first is cheaper, but the second is more robust and bees pass through the holes more easily.

Queen excluders must be treated with care. If the grid is distorted it may let a queen through, and is thus useless. Before this fault is discovered, however, a honey super may be half full of brood.

For European *Apis mellifera* the slots should be 4.14mm wide. (For tropical *Apis mellifera* coffee screen can be used.)

Excluders may be purchased as follows:

plastic sheets, 42.5 × 51.0cm, said to fit 'all 10-frame hives', thickness 0.8mm:
C. ICKOWICZ
Quartier Saint-Blaise
84500 Bollène
FRANCE

metal sheets, many suppliers including:
STEFAN PUFF GmbH
Neuholdaugasse, 8011 Graz
AUSTRIA

framed wire grids, for example:
B.J. ENGINEERING
Swallow Ridge, Hatfield
Norton, Worcester WR5 2PZ
U.K.

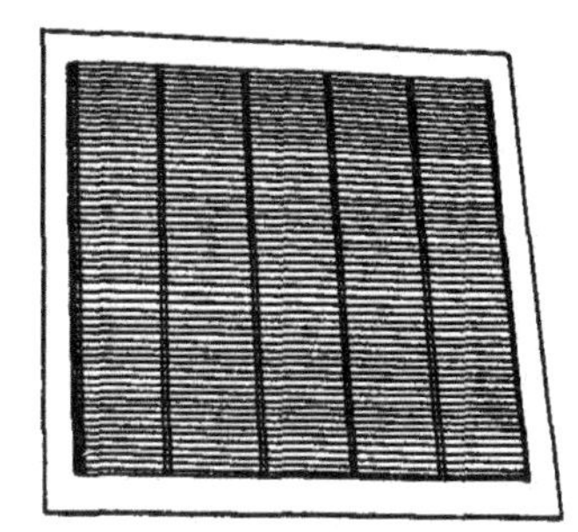

BEE ESCAPE BOARD

A 'bee escape' board (left) is the same size as the cross-section of the hive. It is placed on the hive *below* honey supers that are to be removed to harvest the honey, and it contains a device which ensures that worker bees will pass from boxes above it to those below, but not vice versa, so that the supers can be removed empty of bees. Different devices suit different circumstances — according to whether speed of action, certainty that the device will not be blocked by bees, or some other factor is the prime consideration. Any device relying on a spring mechanism can become ineffective if the spring becomes distorted. Nevertheless the Porter bee escape of this type (illustrated right, with the upper part slid back) is the one most commonly sold. The bees 'escape' from above to below by pushing through the gap between two very light springs, but they are unable to return. Multiple Porter escapes are available.

Bees will usually pass through an escape board between one day and the next. They take less time if there are multiple exits. In cool weather they are slow to move. It is *essential* that all honey supers above the escape board are beetight, or they will quickly be emptied by bees from other hives.

An escape board with no moving parts is preferred by many. It incorporates holes so shaped and positioned that bees will enter from above (and so 'escape') but do not enter them from below, to return. Such conical escape boards are sold by:

A.I. ROOT COMPANY
P.O. Box 706
623 W. Liberty Street
Medina, OH 44258
U.S.A.

Porter escapes are available from:

GENERAL SUPPLIERS

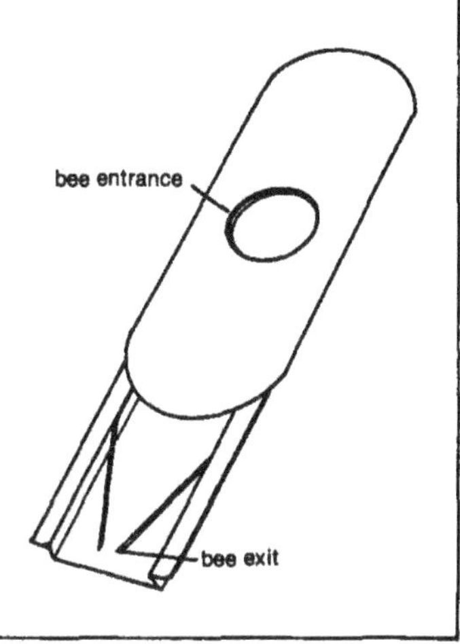

FEEDERS

In most parts of the world it is necessary to make provision for feeding sugar syrup to bees — for instance to counteract some unexpected adverse weather, to build up a small nucleus made in order to increase the number of colonies, or to encourage a swarm, or other bees newly put into a hive, to remain there.

Feeders are placed: at the top of the hive above the top box occupied by bees; or inside the hive with the bees (in the form of a 'dummy' frame); or outside the hive with an entrance only from inside. Larger feeders are preferable for food that the bees must store for a dearth period. When it is important that the bees take the syrup immediately, a 'dummy' frame type is good.

Other materials are used for feeding bees, apart from honey of which an adequate supply should be left in hives as a matter of course. Combs of honey from another source may be used in the hive, but it must be checked that they do not come from a diseased colony. Dry sugar can be fed instead of syrup in warm weather, spread over the inner cover of the hive; it needs no special feeder. Dry sugar feeding will not lead to robbing, which syrup or honey feeding can do if other bees have any access to the food (most usual with an outside feeder).

Pollen and/or pollen substitute is fed by beekeepers in some areas where pollen supplies are deficient, but this also needs no special feeder.

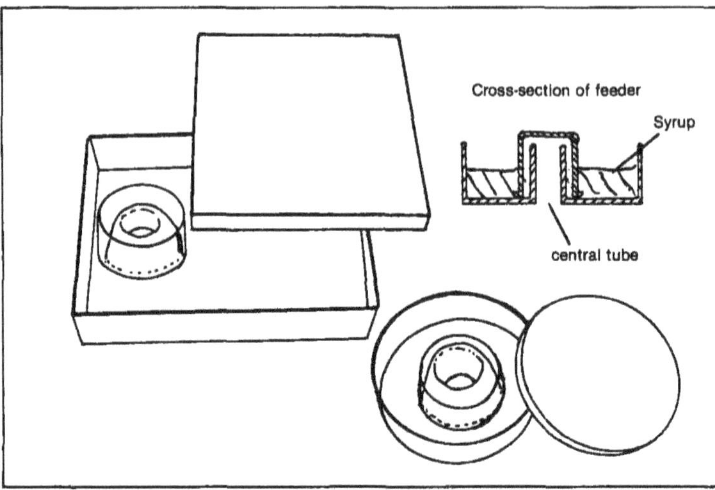

FLAT TOP FEEDER

The drawings on the left show a large square feeder (the size of the hive cross-section) and also a smaller round feeder. Both have a central tube through which the bees enter from below, and a provision for them to have access to the syrup by walking down the roughened outside of the tube. An outer cylinder closed at the top prevents them from getting access to, or drowning in, the bulk of syrup.

The feeders shown are of plastic: the square one fitting the Dadant-Blatt hive is from:

LEGA SDF
Via Armandi 19, 48018 Faenza
ITALY

Round feeders are widely available.
A square plastic feeder of intermediate size can be obtained from:

R. LORHO
Saint-Loup, 28360 Dammarie
FRANCE

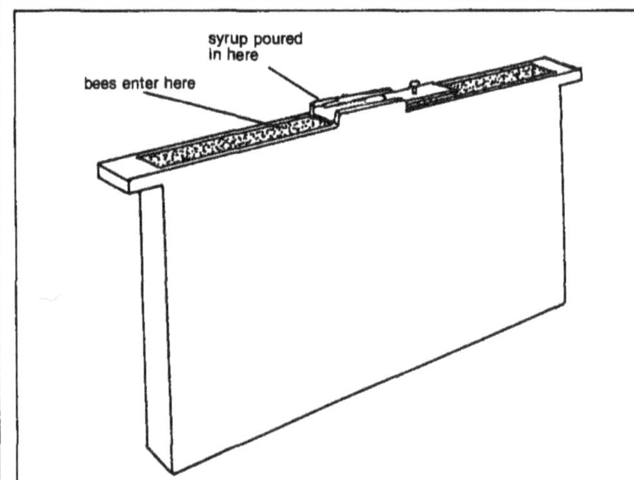

DUMMY FRAME FEEDER

This is sometimes called a division-board feeder. It conforms to the size of the frame-plus-comb in the brood box, where it replaces a complete frame. The bees enter from the top, and inside there is a float or some other provision to protect the bees from drowning in the bulk of the syrup. This feeder is safe from robbing, and its contents are quickly available to the bees, but the hive must be opened to fill it. The feeder shown is of plastic, for Langstroth hives, and is sold by:

OLIFIN PRODUCTS
P.O. Box 10217, Te Rapa
NEW ZEALAND

Another feeder of a similar type is obtainable from:
STANDARD
Kifisias 75, N-Iraklion
Athens
GREECE

PLASTIC PAIL

The plastic pail shown below has a tightly fitting press-on lid, with a fine-mesh insert at the centre. It is filled with syrup (4.5 litres) and inverted over a feed hole in the top cover of the hive.

The pail shown is sold by:
A.I. ROOT COMPANY
P.O. Box 706
623 W. Liberty Street
Medina, OH 44258
U.S.A.

Many beekeepers use, instead of a special feeder, an inverted friction-top metal honey tin/can/pail that holds 2kg or 5kg of syrup, and these can be purchased from almost any beekeeping equipment supplier. Two or three holes are punched at the centre of the lid, using a nail (jagged side inside the pail). More holes can be used, but if they extend all over the lid, the syrup is liable to leak out when the can is inverted over the colony.

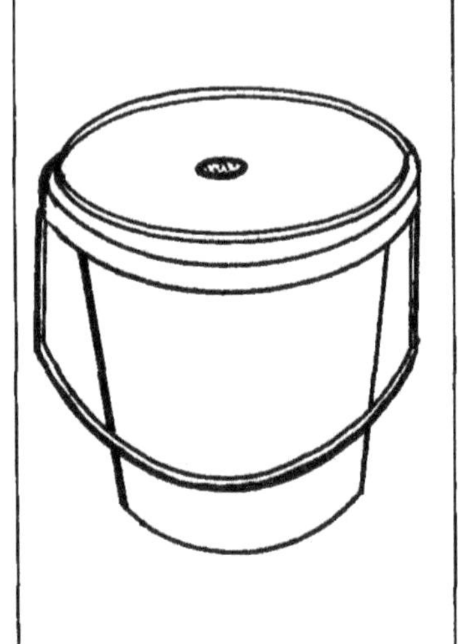

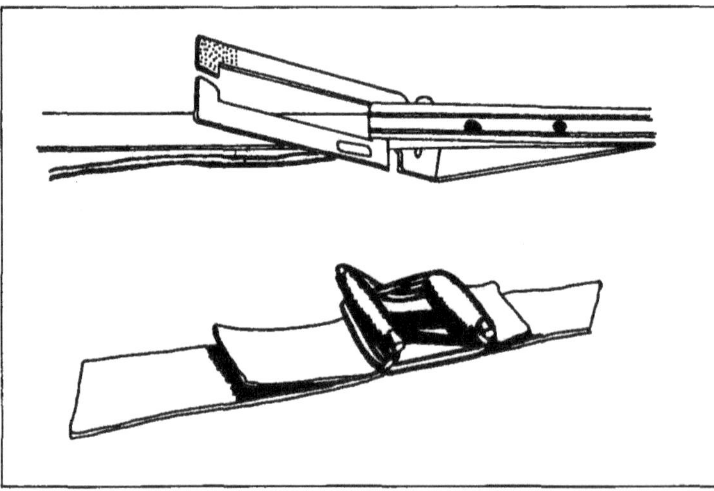

DEVICES FOR SECURING HIVES FOR TRANSPORT

Frame hives are often transported by truck from one honey flow to another. It is essential that all hive boxes, and cover and floorboard, are fastened together so that they do not slip apart and let bees escape. Ventilation for the bees is also essential, and the lid of the hive is usually replaced by a perforated metal screen. The drawing shows two types of buckles for rayon/nylon straps to put round the hive; styles sold by general suppliers vary from country to country.

Apart from tightly secured straps, various metal devices are sold for permanent fitting to the hive boxes, with a closure to be applied before moving which locks the hive together; not all are suitable for large-scale operation. Large metal pointed staples are sometimes knocked into the boxes, but they cause damage and are not be be recommended.

Available from:
GENERAL SUPPLIERS

HIVE TOOL

The name 'hive tool' is given to a strong metal bar (usually of high quality spring steel) about 20-25cm long, which is shaped at the two ends in special ways. One end, often bent at right angles to the bar itself, is broad and with a sharp edge; it is used for scraping wax or propolis (a sticky resin bees use as a building material) off a wooden surface. For good leverage in loosening e.g. frames or top-bars, the other end is made narrower, or it may be specially shaped, for instance as shown in the upper drawing. A hive tool is useful for separating hive boxes, and is kept constantly at hand when inspecting hives. Any general beekeeping supplier sells hive tools, and it is worth trying several in your hand to see which suits you. The two on the left are made by:

MAXANT INDUSTRIES INC.
P.O. Box 454
Ayer, MA 01432
U.S.A.

FOUNDATION AND COMB

This section covers both comb foundation and the equipment for making it. The cell size is critical for both foundation and comb, so different types of bees are dealt with separately.

Foundation is made of beeswax. A mixture of other waxes presents problems when combs are finally melted down and the beeswax recovered. For the use of plastics, see plastic frames (below right).

The diagram below shows (enlarged) the pattern of hexagons pressed into a flat sheet of beeswax when it is made into comb foundation.

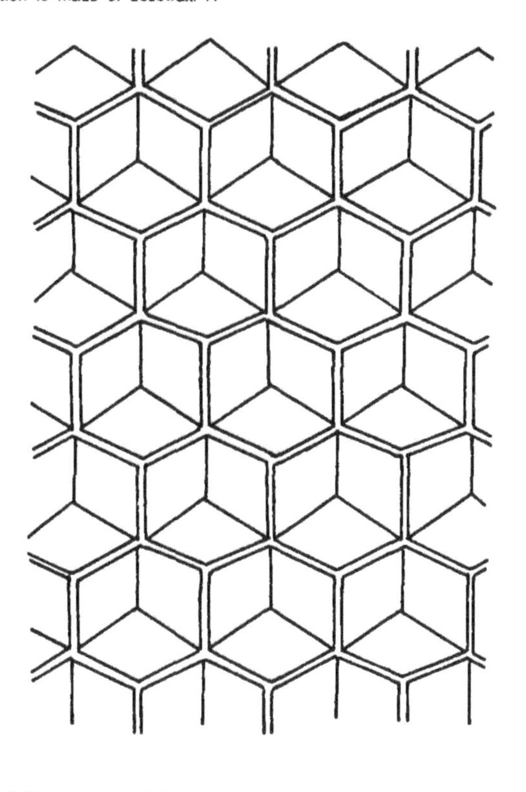

BEESWAX FOUNDATION FOR EUROPEAN 'APIS MELLIFERA'

This is obtainable from almost any general beekeeping supplier where *Apis mellifera* is used. In North America most foundation is sold with strengthening wires embedded in it; elsewhere some beekeepers embed the wires themselves, securing them to the frame in the process. In Asia, both worker and drone foundation is manufactured by:

IMELDA'S BEEKEEPER SUPPLIES
1910 F. Tirona Benitez Street
Malate, Manila
PHILIPPINES

FOUNDATION DIES

These provide the least expensive way of making foundation on a small scale. Molten beeswax is poured into a 'forming tray' to form a thin sheet, and the sheet is laid between a pair of matched dies (plastic sheets embossed with the hexagon shape of cell bases, shown on the left). The 'sandwich' is passed through a wringer or under a heavy roller. Alternatively, an oil drum filled with wet sand, or sand and water, can be rolled over the 'sandwich' which has been laid carefully on a flat board. The hinged plastic dies are sold in two sizes, 28 × 43cm for Dadant frames and 23 × 43cm for others, with cell size for worker or drone. The tray is manufactured by:

H.T. HERRING & SON
14 Severn Gardens, East Oakley
Basingstoke, Hants. RG23 7AT
U.K.

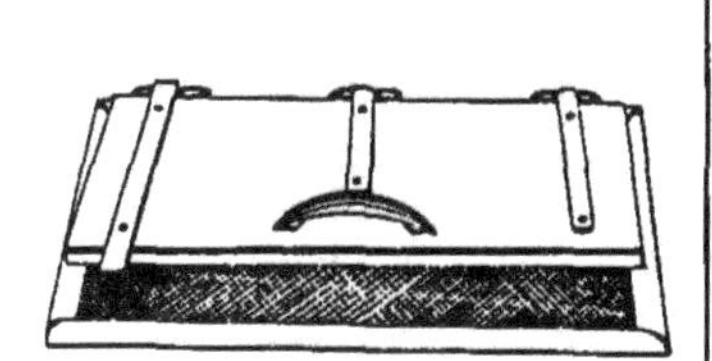

MOULD FOR MAKING EUROPEAN 'APIS MELLIFERA' FOUNDATION

The mould for making European *Apis mellifera* foundation is heavier than the embossed plastic dies (see above) but is more straightforward to use. Molten beeswax is poured onto the lower die, which constitutes the base of the tray, and the hinged lid (upper die) is closed on to it. One manufacturer of such a mould is:

LEAF PRODUCTS
24 Acton Road, Long Eaton
Nottingham NG10 1FR
U.K.

ROLLERS FOR MAKING EUROPEAN 'APIS MELLIFERA' FOUNDATION

A previously formed long strip of beeswax sheet is passed between two rollers embossed to serve as dies for the foundation. (The long strip has been obtained by using a pair of plain rollers).

Three suppliers are:
ROGER DELON
83 Route de Grand-Charmont
25200 Montbéliard, (Doubs)
FRANCE

CHR. GRAZE KG.
Strümpfelbacherstraße 21
7056 Weinstadt 2, (Endersbach)
W. GERMANY

BERNHARD RIETSCHE
Bienengerätefabrik
7616 Biberach/Baden
W. GERMANY

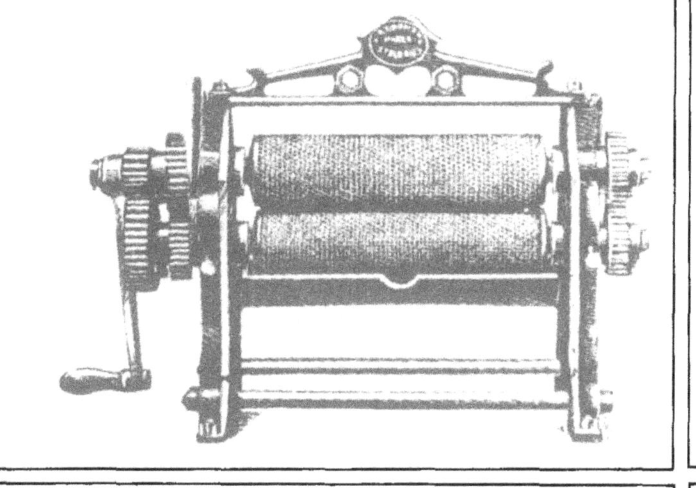

PLASTIC FRAMES WITH INTEGRAL COMB (FOR 'APIS MELLIFERA')

Many plastic frames with integral plastic foundation (or alternatively comb) are produced. They are suitable for large-scale beekeepers, and benefits include greater strength when extracting honey at high speed, ease of sterilization, and saving of time used in assembling wooden frames.

In many conditions bees seem to prefer beeswax to plastic in comb or foundation, when both are used. If plastic is tried, it should be during a good honey flow, and *all* frames in a super should be plastic. Frames used successfully, with integral foundation on which the bees build combs are Pierce Plastic Frames, made by;

PIERCO INC.
1495 W. 9th Building
501 Upland, CA 91786
U.S.A.

BEESWAX FOUNDATION FOR TROPICAL AFRICAN 'APIS MELLIFERA'

The cell size for these bees is quoted as 1050 cells/sq dm. The following firm is believed to supply suitable foundation to this specification:

JOHN RAU & COMPANY (PVT.) LTD.
2 Moffat Street
P.O. Box 2893, Harare
ZIMBABWE

FOUNDATION MOULD
Manufactured by:
LEAF PRODUCTS
24 Acton Road, Long Eaton
Nottingham NG10 1FR
U.K.

FOUNDATION ROLLERS
Manufactured by:
TOM INDUSTRIES
P.O. Box 800
El Cajon, CA 92022
U.S.A.

BEESWAX FOUNDATION FOR 'APIS CERANA'

The following firms are believed to supply suitable foundation:

IMELDA'S BEEKEEPER SUPPLIES
1910 F. Tirona Benitez Street
Malate, Manila
PHILIPPINES

RAWAT APIARIES (Himalayas)
Ranikhet, Dist. Almore, UP
INDIA

FOUNDATION DIES
No foundation dies or moulds are known to be on sale. Matched dies like those for *Apis mellifera* (see above) would help many beekeepers in Asia. The initial production of the form from which dies are made is very expensive, and the manufacturer would need either to be assured of a large number of orders, or to receive some financial support, before he or she could produce these dies.

FOUNDATION ROLLERS
Manufactured by:
TOM INDUSTRIES
P.O. Box 800
El Cajon, CA 92022
U.S.A.

BEESWAX FOUNDATION FOR 'APIS FLOREA'

The firm below made (to order) rollers for *Apis florea* foundation, to fit their own foundation mill. They also supplied wax-melting tanks, and rollers for making the preliminary wax sheets.

LOTLIKAR AND SONS
A-1/4 Pioneer Co-op Society
Panvel 410206, Kulaba M.S.
INDIA

QUEEN REARING

Queen rearing can be a profitable undertaking, both in financial terms and in improvement of colony performance and ease of handling. It must be done to a strict pre-planned timetable. Since queens mate in the air, no control over the male line is possible, except in isolated mating apiaries, or by using instrumental insemination. All equipment here is for European *Apis mellifera*, but most could be used or adapted for other bees.

One specialist firm is:
CHRISTIAN NICOT
Maisod, 39260 Moirans-en-Montaigne
FRANCE

GRAFTING TOOL

For transferring a young larva into a cell cup (see right), operators may use something to hand, for instance a pointed piece of yucca or other leaf. Alternatively they may prefer a specially shaped blunt 'needle' with a handle that allows them to work comfortably but with care and precision.
Grafting requires patience, good eyesight, and steady hands, but no booklearning.

Available from:
GENERAL SUPPLIERS

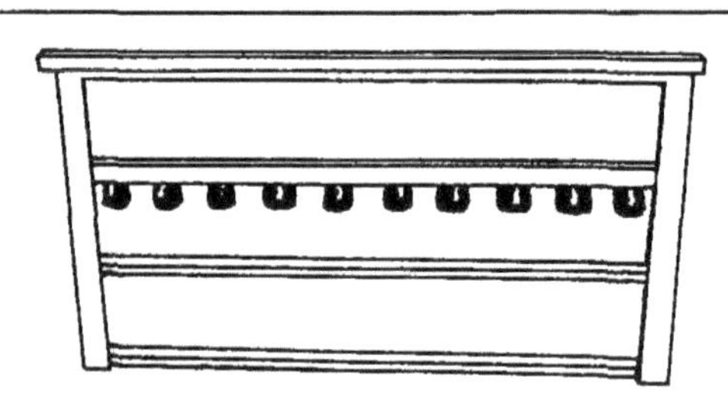

CELL CUPS AND FRAME

The larvae are grafted (see tool left) into artificial 'cell cups' (right) with a supply of royal jelly, and the cups are often mounted on 3 horizontal bars fixed between the end bars of a frame which contains no foundation, 10 to each. The frame is inserted in a strong queenless colony of bees.

GENERAL SUPPLIERS

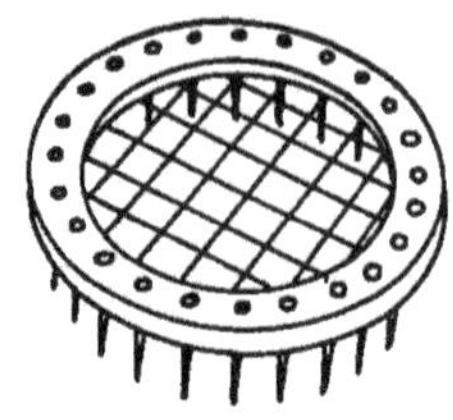

QUEEN MARKING

A queen bee may be marked for identification by gently placing the fine threads of a press-in cage over her on the comb. A spot of quick-drying paint is then placed on the queen's thorax. Or tiny coloured discs may be stuck on; these are either numbered, or are made in the 5 colours of the international code for queen marking.

Available from:
GENERAL SUPPLIERS

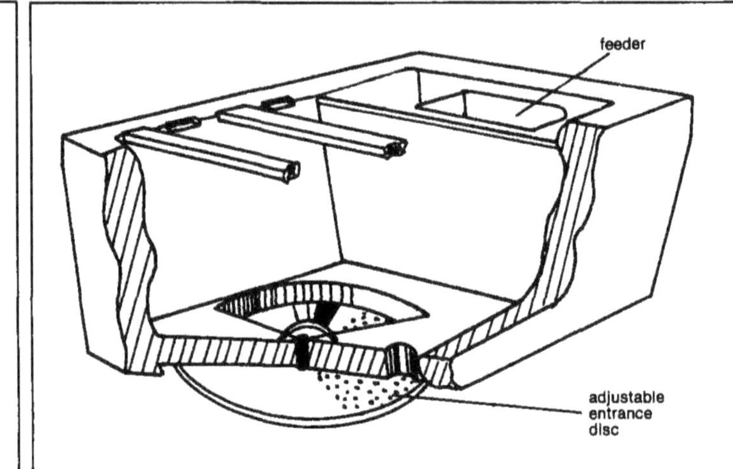

MATING HIVE

After the queen cells are sealed, each is put into a small mating hive. The queen will shortly emerge and fly out to mate with drones in the air when a few days old. The mating hive can be like part of a brood box, containing 2 or 3 frames. Or it can be a tiny hive (as illustrated) which uses fewer bees. Queens are likely to mate more quickly from a small hive. Such hives are often of polystyrene, which provides better heat insulation than wood. They contain 4 top-bars from which the bees build sufficient comb to allow the queen to start laying. A feeder is provided. The model shown is fitted with an adjustable entrance at the bottom, and may be suspended for safety. Mating hives can be purchased from many suppliers; illustrated is that from:

R. STEELE AND BRODIE
Stevens Drove, Houghton
Stockbridge, Hants SO20 6LP
U.K.

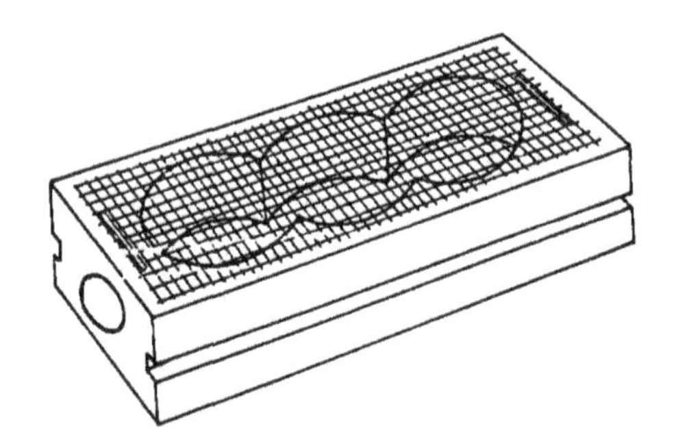

QUEEN MAILING CAGE

The queen rearer uses this type of cage for sending queens to the purchasers. The cages are made by drilling into a small solid block of wood, and one hole is filled with candy to provide food for the queen and attendant workers in transit. One side of the cage is of wire mesh to allow ventilation.
Available from:
GENERAL SUPPLIERS

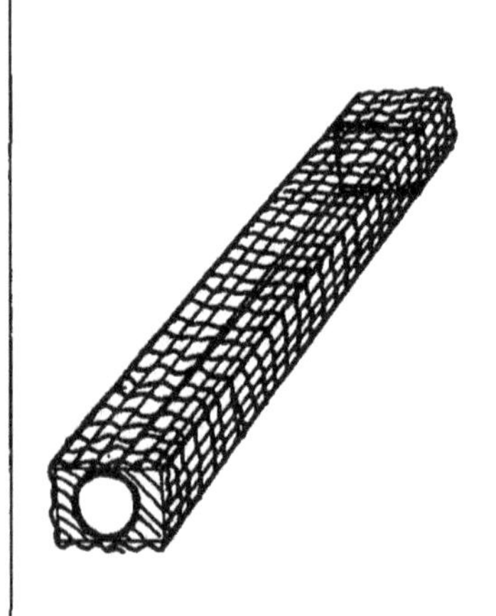

INTRODUCING CAGE

The introducing cage is thin, to be slipped into the space between two frames. Dimensions of the cage shown are 89 × 19 × 13mm: the wire gauze leaves holes about 3mm across. These are as wide as possible because workers of the new colony must have direct physical contact with the queen in the cage, so that pheromones* can pass between them. It is then more certain that she will be accepted by the colony. The queen is inserted into this cage without food, and the bees feed her through the gauze. In some types of cage the mesh is too small. (The workers that travelled with her are removed and killed beforehand; they can be of no benefit, and might possibly carry a new disease infection into the colony.)

*Substance produced by insects and animals for detection and response by others of same species.

Available from:
GENERAL SUPPLIERS

APPARATUS FOR INSTRUMENTAL INSEMINATION

Methods for insemination of the queen with drone semen have been so well developed that they are now routine in some countries. But the apparatus required is expensive, and it is a waste of time and money to purchase it unless (a) a properly controlled selection programme can be used for bee breeding, and (b) it will be more beneficial to use instrumentally inseminated queens than naturally mated ones. The apparatus is not sold by many general beekeepers' suppliers, but the first address listed is one of the few that does and they also sell an excellent set of 111 full-colour 35mm slides with a printed manual based on them, taking the operator through every stage of the process.

DADANT AND SONS INC.
Hamilton, IL 62341
U.S.A.

APISMAR
Calle 40 no.492
La Plata, B.A.
ARGENTINA

CHIRANA, EXPORT-IMPORT
92175 Piestany
CZECHOSLOVAKIA

J. HAIDINGER
Finkenweg 5, 8042 Schleissheim
W. GERMANY

HESSISCHE LANDESANSTALT FÜR LEISTUNGSPRÜFUNG
Aussenstelle für Bienenzucht
Erlenstraße 9, 3575 Kirchhain
W. GERMANY

W. SEIP
Hauptstraße 34-36
6308 Butzbach-Ebersgöns
W. GERMANY

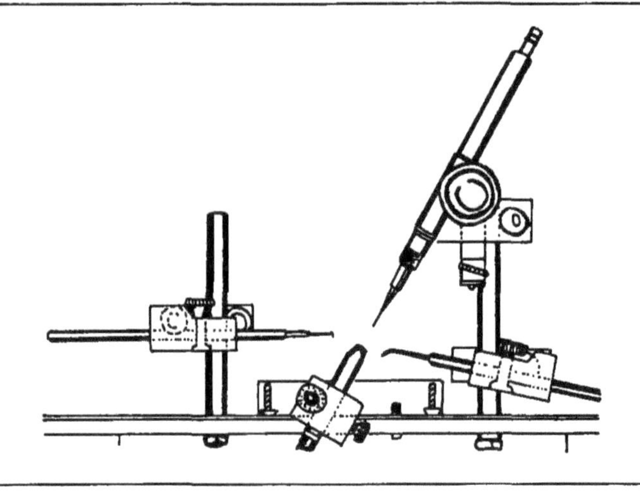

REMOVING HONEY FROM THE HIVE

The bees must first be made to leave the honey supers. This can be achieved by brushing and shaking them off the individual combs, or by using an 'escape board' (see Hive Fittings, p.223). Alternatively the bees can be driven from the supers down into the brood chamber by using a 'fume board' to apply a bee repellent, or the supers can be taken off the hive and the bees blown out with a 'bee blower'.

An escape board involves two visits to the hive, one to insert it and another to check that the supers are bee-free, and then to remove them. The other methods need only one visit, but they disturb the bees more. It is therefore best to work late in the day, when flight activity is decreasing, to reduce the chance of subsequent robbing of honey by bees from other hives. If robbing seems to be occurring, it can be helpful to apply a fine spray of water to the hive fronts, or wherever bees are congregating — by reducing the temperature this makes the bees less active.

FUME BOARD

This is a shallow box or tray made of insulating board such as pressed fibre — of the same cross-section as the hive. The internal depth of the tray is important and depends on the volatility of the repellent sprinkled on it. At temperatures above 27°C an insulated cover is helpful, to prevent heat from the sun vaporizing too much of the repellent. For benzaldehyde (artificial oil of almonds) a depth of 5cm has been recommended. After sprinkling benzaldehyde on the inside of the fume board, this is inverted over the (uncovered) top super; a white cloth on top of the board will prevent too rapid evaporation. After a few minutes the bees should have left the top (shallow) super; benzaldehyde is not effective with a full-depth hive box. The super is then removed, and if necessary the fume board is placed similarly on the one below, and so on.

If many supers are to be removed from a heavily populated hive (and especially if some of the honey is unsealed), increasing difficulty may be encountered with successive supers, since more and more bees are being crowded into a smaller space. (One commonly held objection to the use of any repellent is possible contamination of the honey, but this applies much less to benzaldehyde, which is used as a food flavouring, than to carbolic acid which was used earlier as a bee repellent.)

Available from:
GENERAL SUPPLIERS

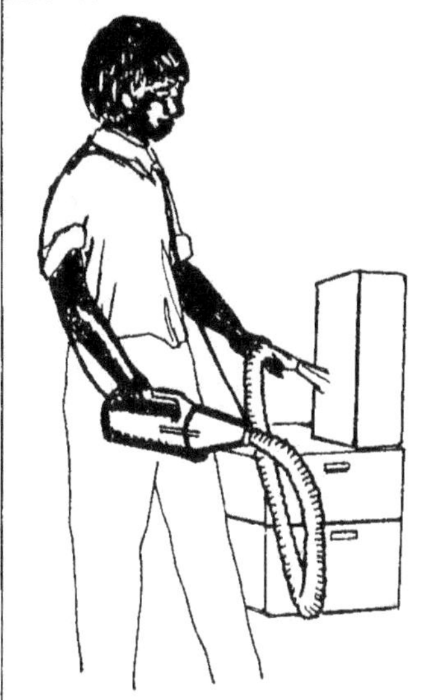

BEE BLOWER

Unlike smoke or a bee repellent, an airstream used to blow bees out of a hive does not introduce any possibility of honey contamination.

A bee blower is normally powered from an electricity supply, but one could be devised to be operated by some other form of power. It is rather like a vacuum cleaner in reverse, and is used by standing a super (open above and below) either on top of the hive (as shown) or on a stand constructed like a sawing horse, placed just in front of the hive. The bees blown out of the super find their way back to the hive entrance. The blower shown operates on 110V, and is made by a specialist firm:

SOUTHWESTERN OHIO HIVE PARTS CO.
Monroe, 629 Lebanon Street
OH 45050
U.S.A.

UNCAPPING HONEY COMBS

Framed combs of honey taken from the hive must be uncapped (to remove the wax seals) before they are put in an extractor (see below).

UNCAPPING KNIFE

The knife shown here is of a standard type. The essential features are (a) that the knife is longer than the depth of the frame, and (b) that the handle is offset for convenience in use. The knife can be heated by standing it in hot water. Often two knives are used, one being heated while the other is used.
Available from:

GENERAL SUPPLIERS

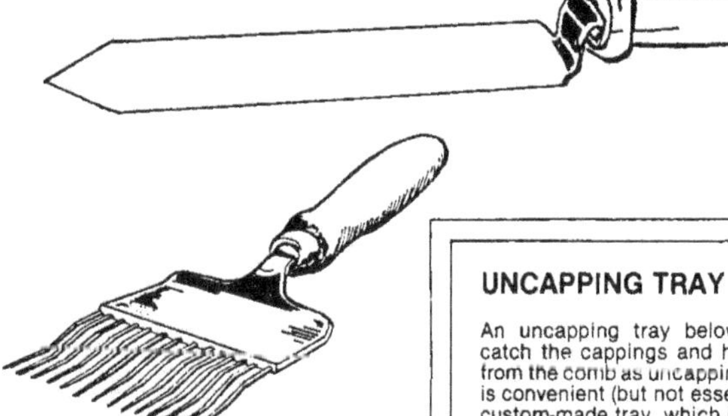

UNCAPPING FORK

These are of various widths, and many of them are narrower than the depth of a comb. They are operated by sliding the fork under the cappings from one end of the comb to the other. The narrow forks are useful when the shape or surface of the comb is irregular. The fork in the drawing (above), with offset tines, is made by:

BLOSSOMTIME
P.O. Box 1015
Tempe, AZ 85281
U.S.A.

UNCAPPING TRAY

An uncapping tray below is used to catch the cappings and honey that fall from the comb as uncapping proceeds. It is convenient (but not essential) to use a custom-made tray, which has a sheet of wire mesh near the bottom through which the cappings drain. The frame is held firm by an upward pointing projection on the cross bar. General suppliers sell various types of uncapping tray, some with additional features, which suit individual preferences.

Available from:
GENERAL SUPPLIERS

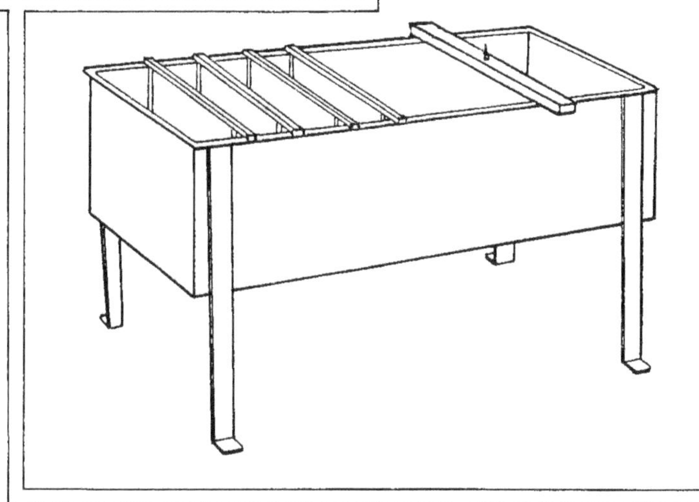

HONEY EXTRACTORS

These operate by centrifuging the honey out of the combs. The extractor is a cylindrical container with a centrally-mounted fitting that supports combs or frames of uncapped honey, and a mechanism that rotates the fitting (and the combs) at speed. The honey is thrown out by centrifugal force to the inner wall of the extractor, whence it falls by gravity to the bottom. Very near the bottom a honey gate (see over) is fitted, allowing the honey to be drained out when required. A free space is left below the frames so that a certain amount of honey can accumulate, but honey must be drained off before it reaches the supports of the frames.

The temperature of the room is very important when extracting honey from the combs, because honey flows very much more quickly when warm than when cold, and less is left in the cells. Also, if high speeds have to be used to force the honey out, the combs are more liable to break. With very high-speed (electrically operated) extractors, the speed has to be increased gradually to prevent damage to combs.

TANGENTIAL EXTRACTOR

This was the first type to be developed, and is still much used, especially in small-scale beekeeping. The axis is vertical, and framed combs (often 2, 3 or 4) are supported in baskets, or against vertical grids, arranged tangentially, i.e. at right angles to the radius. The frames must normally be spun twice, once with each of the two sides outermost. Some tangential extractors are self-reversing. These extractors can be obtained from almost any general supplier. A firm that specializes in well made honey extractors is:

MAXANT INDUSTRIES INC.
P.O. Box 454
Ayer, MA 01432
U.S.A.

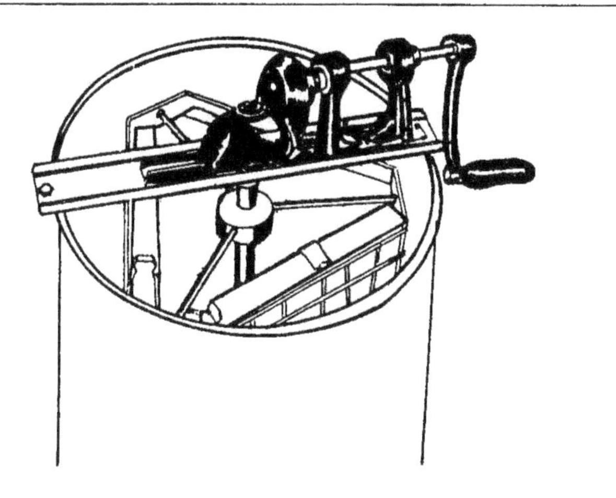

HONEY PROCESSING

All previous operations have been on bees or hives. When starting to do work on honey, it is important to remember that honey is a food, and that appropriate standards of hygiene must be maintained. Also, bees not only make honey; they quickly get the scent of any honey left unguarded, and collect it to take back to their hives. This can happen in an astonishingly short time. So from the moment the honey combs free from bees are taken off the hive, *they must be in a beetight building or enclosure.*

Uncapping and extracting (by whatever method) must be done in a room that allows no access to bees. A few bees may be brought in on clothes or on combs, so vents that allow bees to fly out of the room but not to re-enter are helpful. There is one exception to the above rule: if a good honey flow is still in progress, the bees may continue to work it and ignore the honey being dealt with.

Bees in the honey house are objectionable from a hygienic point of view, because when they fly round trying to escape, they release excreta on to walls and floor and this is unacceptable in a food-processing area.

In the simplest operation, pieces of honeycomb are placed in a cloth, which is hung up and left for the honey to drip out (this is 'run honey', the best), and then squeezed to force out as much as possible of the remaining honey.

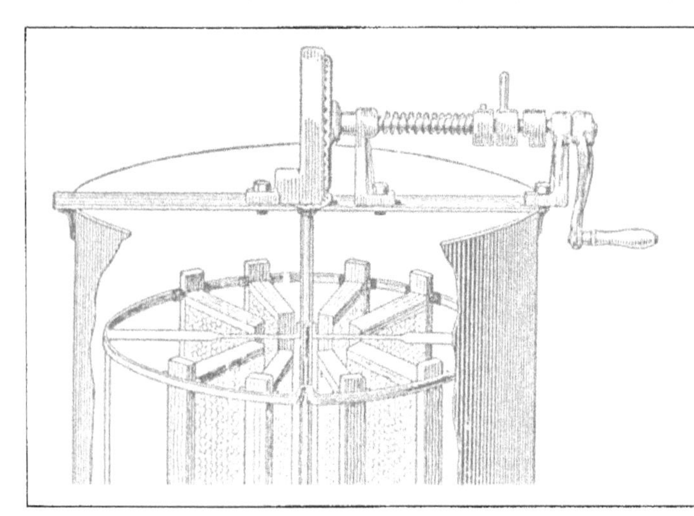

RADIAL EXTRACTOR

A radial honey extractor is like a tangential one, except that the frames are placed radially. The cylinder is larger, and is often made to hold the frames from one honey super (9 or 10) or a multiple of this number. Frames are placed with the top-bar outwards. More power is needed to operate these extractors, and large ones are electrically operated, but a 9 or 10 frame extractor can be operated by hand, or by foot using an adapted bicycle mechanism. Only one spinning is needed.

Available from:
GENERAL SUPPLIERS

HONEY GATE

Taps designed for water flow cannot be used for honey, which is very viscous and flows only slowly. Honey gates for fitting to extractors and tanks should be hygienic and easily cleaned; they should cut off the flow of honey instantly, with no drip, as soon as they are closed, and they must incorporate a safety device (often a screw) to prevent accidental opening from the closed position.

Honey gates are usually made of brass, stainless steel or plastic, and the diameter of the opening can range from 32mm to 76mm. Most large beekeeping suppliers sell them.

Available from:
GENERAL SUPPLIERS

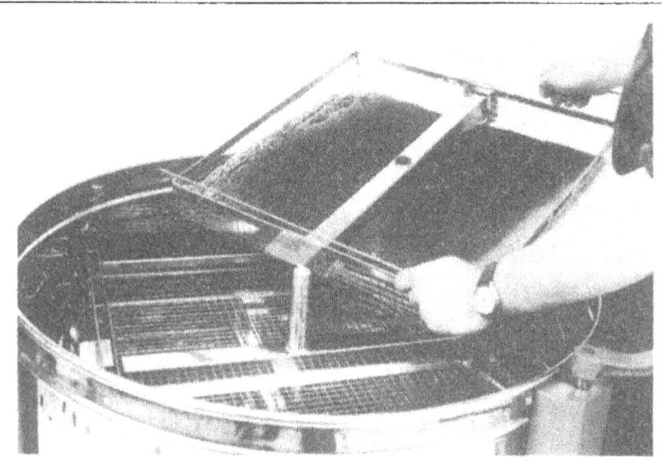

EXTRACTOR FOR COMBS FROM TOP-BAR HIVES

These combs do not have the support of a full frame, or the strength of combs built on wired foundation. They cannot withstand the force of a normal extractor, but a small tangential extractor can be adapted by providing wire-mesh baskets in place of the usual grids. Unframed combs (or pieces of comb) are carefully placed in the baskets. The extractor must be spun twice, once with each side of the basket innermost.

A larger extractor (illustrated), for combs from a top-bar hive, contains 6 pairs of baskets mounted horizontally, the whole taking 12 combs. Baskets are removed from the top, pair by pair, and combs inserted before they are replaced in the extractor. This extractor is wired for electric operation, but could be adapted for use without electricity. It is produced by:

ETS THOMAS FILS SA
65 rue Abbé Georges Thomas
BP No. 2, 45450 Fay-aux-Loges
FRANCE

HONEY STRAINERS

Commercially sold honey strainers are designed to take honey as it leaves the extractor, containing no more than small bits of wax from cappings. Modern quality requirements demand that final straining is through a very fine mesh, and this process is speeded up if all but the smallest particles have been removed first, by one or more strainers of larger mesh. As with other operations on honey, straining is faster if the honey is warm; honey flows roughly twice as fast for every rise of 10°C. Light and dark combs should be strained separately, into different containers, since the flavour of the darker honey may not be as fine.

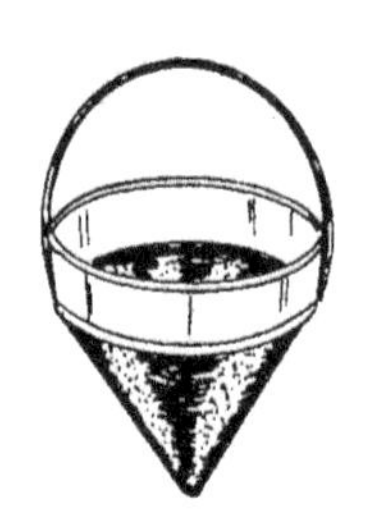

SIMPLE HONEY STRAINER

The least expensive type of honey strainer is suspended from the honey gate of the extractor. On the left the honey is strained through a single cone of finely perforated metal. On the right an upper coarse wire mesh retains the larger particles, speeding up the flow through the lower fine wire cloth; below right these two strainers are shown separately. The strainers shown are sold by:

E.H. THORNE (BEEHIVES) LTD
Wragby, Lincoln
LN3 5LA
U.K.

The above strainers are practical only for small amounts of honey. Somewhat similar strainers can be purchased with (or to fit at the top of) a polythene tank holding 70kg (see below).

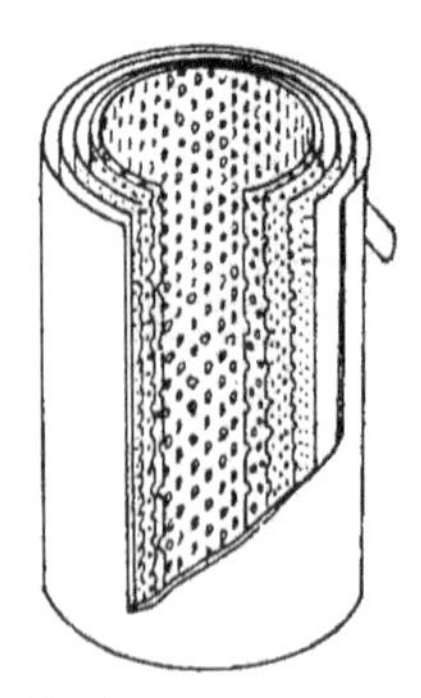

Dimensions in cm:
External diameter: 43
Diameters of screens: 18, 23, 28, 33
Height: 76
Outlet diameter: 7.5

OAC HONEY STRAINER

This well tried strainer for larger-scale operations was developed at Ontario Agricultural College (now the University of Guelph), in Canada. It consists of a cylindrical tank with a series of four cylindrical coaxial screens, one inside the other. From the centre, they have approx. 5, 12, 20, 30 mesh/cm. Honey enters the tank at the top, inside the innermost screen, with the largest mesh. It passes through the screens and is drawn off, also near the top. Each screen, and the tank, has a drainage gate at the bottom. This strainer has a large straining area for each mesh, and will handle 2 tonnes of honey a day at 30°C. In temperate climates, it can be operated successfully without heating the honey. The OAC strainer is sold by:

F.W. JONES & SON LTD
44 Dutch Street
Bedford, Quebec JOJ 1AO
CANADA

HONEY STRAINER AND SETTLING TANK

For small-scale operation, many beekeepers use a tank with two or more screens at the top, the upper one having the larger mesh. A honey gate (see above) is fitted near the bottom for removing honey, and the tank can be tilted for final draining. When honey stands in the tank, any remaining particles of beeswax rise to the top (hence the term 'settling' tank.)

Many general suppliers sell these tanks. The tank shown is made of polythene and holds 70kg of honey; it is sold by:

E.H. THORNE (BEEHIVES) LTD
Wragby, Lincoln
LN3 5LA
U.K.

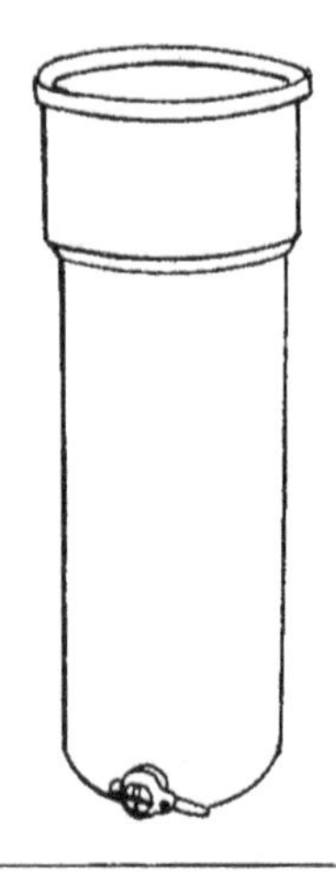

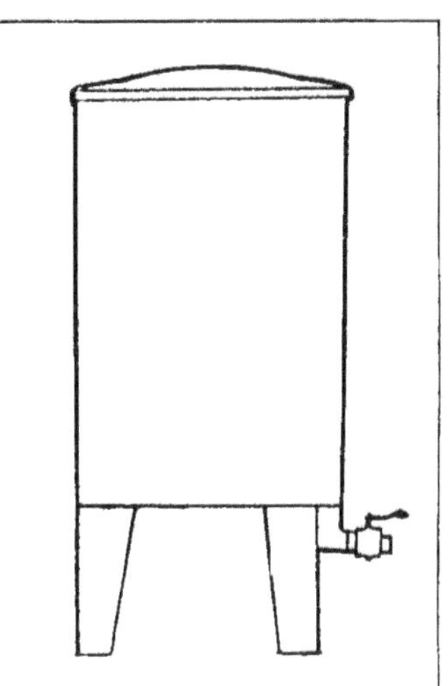

HONEY STORAGE TANKS

Containers in which honey is stored should be inert (giving no interaction between the vessel and the honey) and easy to clean. Honey is a food product, and it has a delicate flavour, and on both counts must not be stored in metal drums that cannot be cleaned, or that are scratched or damaged. Plastic and stainless steel are ideal materials for smaller and larger honey tanks, respectively. The containers must also be tightly closed and moisture resistant, or the honey in them will absorb moisture and may then ferment. The tank shown is made of stainless steel and holds 2 tonnes. It is provided with a stand since it is too heavy to be tilted for emptying. The base slopes down to the gate fitted at the lowest point.

A specialist manufacturer of these tanks is:

MAXANT INDUSTRIES INC.
P.O. Box 454, Ayer, MA 01432
U.S.A.

HONEY CONTAINERS FOR MARKETING

As with honey storage tanks, the material must be inert and easily cleaned. Airtightness is essential, especially in humid climates, or the honey will absorb moisture from the air and will then be liable to ferment. Only plastic containers are described here; there is no chance of contamination as there is with metals (other than stainless steel). Glass jars are heavy, liable to break, and cannot be stacked one inside the other, so they are expensive to transport.

LARGE POLYTHENE PAILS (2 to 30kg)

These usually have a reinforced rim, a press-on lid fitting tightly, and a wire handle. Those illustrated right are sold by:

PRO-WESTERN PLASTICS LIMITED
150 Riel Drive, P.O. Box 261
St. Albert, Alberta, T8N 1N3
CANADA

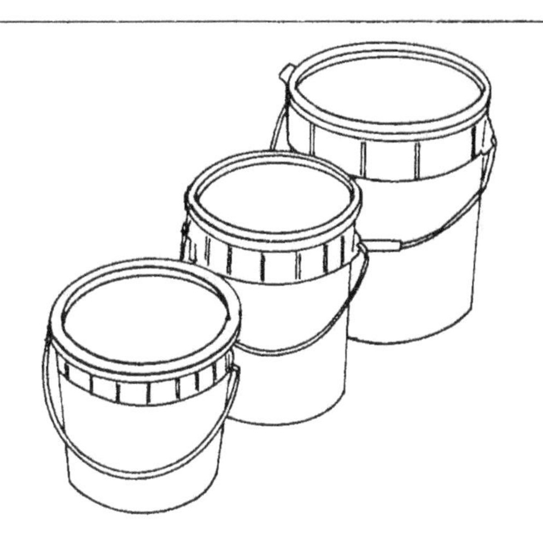

SMALL PLASTIC POTS (0.5 to 2kg)

These have tightly fitting (re-usable) press-on lids. They may be tall or squat, opaque or transparent. For liquid honey, tall transparent pots are preferable, to minimize the chance of leakage and to show the honey off to advantage. For granulated (crystallized) honey, opaque pots are often used, because surface irregularities in crystal formation are then not visible.

Suppliers of tall pots (illustrated right) include:
SAF, s.n.c.
Via Liguria 17, 36015 Schio (VI)
ITALY

Suppliers of squat opaque pots:
LILY CUPS DIVISION
P.O. Box 2195, Auckland
NEW ZEALAND

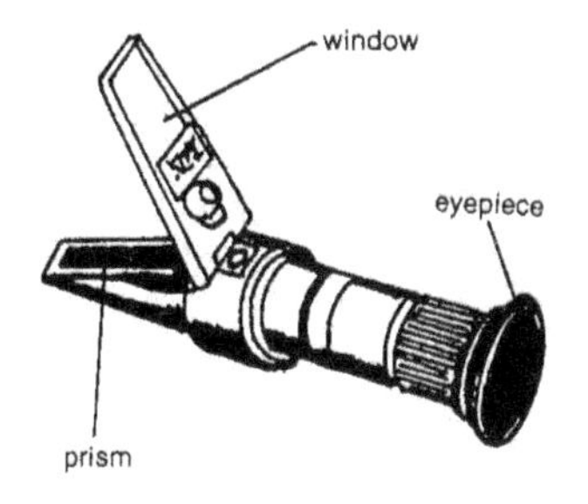

HONEY REFRACTOMETER

A refractometer measures the refractive index of a substance, and one calibrated specifically for honey is very useful because the refractive index depends on the total percentage of sugars in the honey. The refractometer is usually calibrated directly in percentage of water (moisture). The upper limit in the proposed FAO/WHO Codex is 21 per cent, but most honey producers and traders would regard 18 per cent as a proper limit. Instructions are sent out with each refractometer, and they include a table of temperature corrections, since the refractive index is substantially affected by temperature. Some refractometers are calibrated in degrees Brix (a form of measurement of the sugar content in a substance).

In operation, a few drops of honey are placed on the prism (left of drawing), and the hinged window closed down on them, spreading them into a very thin 'sandwich'. On viewing through the eyepiece (right of drawing) against a good light, the scale will be seen, with an indicating line showing the reading. Since so little honey is used, it is most important that it is representative of the sample, and that the glass surfaces are completely dry. Some refractometers incorporate a thermometer. The following firms supply honey refractometers:

BEEMAID
625 Roseberry Street
Winnipeg, Manitoba R3H OT4
CANADA

A. ECROYD AND SON LTD
P.O. Box 5056
25 Sawyers Arms Road
Papanni, Christchurch S5.
NEW ZEALAND

GIFU YOHO CO LTD
Kano-Sakurada-cho 1
Gifu-Shi, Gifu 500-91
JAPAN

STEFAN PUFF GmbH
Neuholdaugasse, 8011 Graz
AUSTRIA

SOPELEM
102 rue Cheptal
92306 Levallois, Perret
FRANCE

PFUND COLOR GRADER

This instrument is used in the world honey market for measuring the 'colour' (darkness) of honey; it compares the opacity of a honey sample with that of a standard 'amber' liquid. The honey sample is placed in a wedge-shaped trough which is moved past a narrow slit in the housing until a colour match is obtained, i.e. the colour density of the honey matches that of the amber wedge. The reading on a millimetre scale is then the 'Pfund scale' reading for the honey, which corresponds to one of the following standardized colour names in the USA (in Canada and Australia the definitions are slightly different): up to 8mm water white; up to 16mm extra white; up to 34mm white; up to 50mm extra light amber; up to 85mm light amber; up to 114mm amber; over 114mm dark amber. Available from:

XPORT
Port Authority Trading Company
1 World Trade Center. 55NE
New York, NY 10048, U.S.A.

POLLEN TRAP

Honey and beeswax are the most commonly harvested hive products. Pollen, compared with honey, has a high protein, vitamin and mineral content, and in some countries is harvested and processed for sale. Harvesting is done by using a pollen trap, a device incorporating a hive entrance in which incoming bees must pass through two parallel grids of suitable mesh, with the result that pollen loads in the bees' hind legs are knocked off and fall into a collecting tray below. It would be dangerous to prevent any pollen entering the hive for more than a day or two, because brood rearing would cease, but use of traps is organised to prevent this.

Most pollen traps on sale are fitted at the bottom of a hive (and must have the same cross-section), either immediately above the normal floorboard or instead of it. Other designs are used at the front of the hive or at the top, with an upper hive entrance. Commercially available pollen traps do not necessarily fulfil all the conditions for successful use under all conditions without harm to the colony. International Bee Research Association Publication M86, *Pollen and its harvesting*, explains the problems and gives recommended designs not on sale.

Pollen is a highly nutritious food, and therefore a good medium for the growth of micro-organisms. For this reason pollen traders in technologically advanced countries may be unwilling to import pollen from untried sources.

The pollen trap shown is sold by:

HONEYBEE PRODUCTS
Amery, WI 54001
U.S.A.

Another supplier is:

KOREA BEEKEEPING APIARY
1155-1, Soong In-Dong
Chongro-Ku, Seoul
KOREA

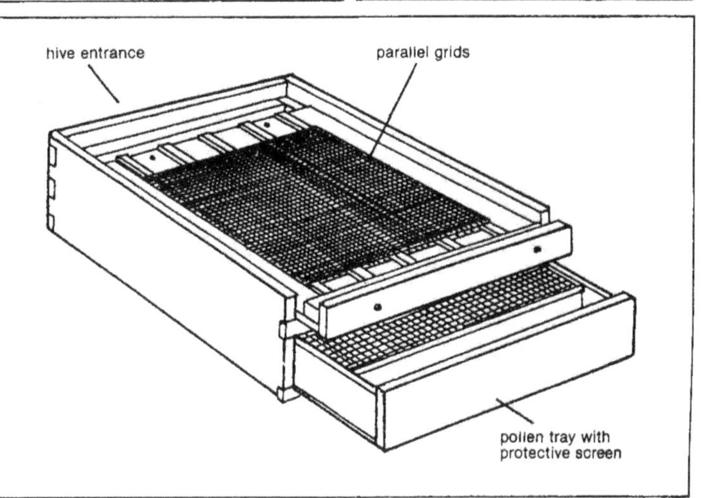

BEESWAX PROCESSING EQUIPMENT

Beeswax is a valuable hive product, and should bring the beekeeper an added income. Unlike honey, it needs no container, and no special care in even long storage. In spite of this it is all too often thrown away. Beeswax has traditionally been exported from tropical countries. There may be little or no local use for it, and a beekeepers' co-operative or similar body may be needed to organize its sale to beeswax traders. When beeswax cappings, combs, etc. have been washed free from honey, or have been cleaned by bees, the beeswax is melted to let everything that is not pure beeswax separate out by sinking to the bottom. It is essential that light combs should be treated separately from dark ones, because light wax will fetch the highest price.

Beeswax *must* be heated in a safe way, or there is danger of a fire. The first apparatus described is ideal in this way, costs nothing to operate, and can produce very high quality wax.

In some areas, traders may be suspicious that blocks of wax offered for sale have stones in them, to add to their weight; any such stones would not be visible. In such areas, it is best to make rather thin blocks, which could not hide stones.

SOLAR WAX EXTRACTOR

The wax pieces are put on a metal base, closed in by four sides and a lid consisting of two sheets of glass 5mm apart; the whole is tilted at a suitable angle to catch the sun's rays. Below the base is an insulating layer to reflect the heat back. Heat is trapped inside the box, and melted wax runs down the sloping base (leaving most of the dross behind), and into a container within the box. A second external container can be incorporated, as in the drawing.

Available from:
GENERAL SUPPLIERS

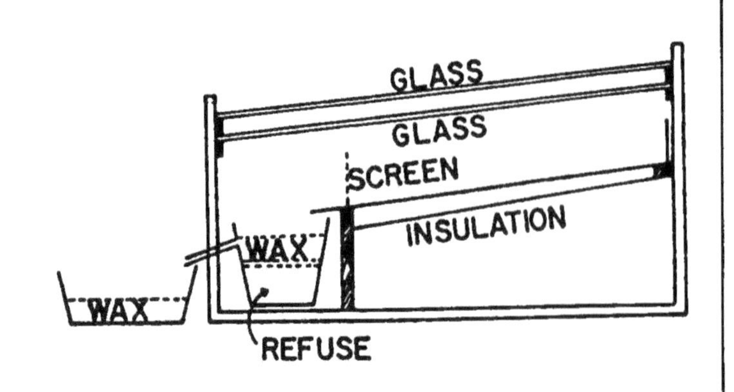

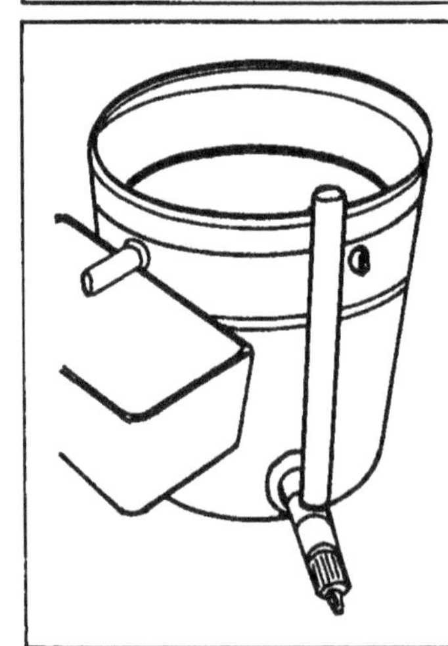

HOT-WATER BEESWAX PROCESSOR

In principle this is a vessel in which water and unprocessed beeswax can be heated together and (a) the beeswax floats on the top and is drained off, leaving behind the dross and the dirty water; (b) more water can be added at the bottom, to raise the beeswax layer to the correct level for draining. The drawing shows a simple type produced by:

HONEY & BEE DIVISION, SHOTS INC.
4418 Josephine Lane
Robbinsdale, MN 55422
U.S.A.

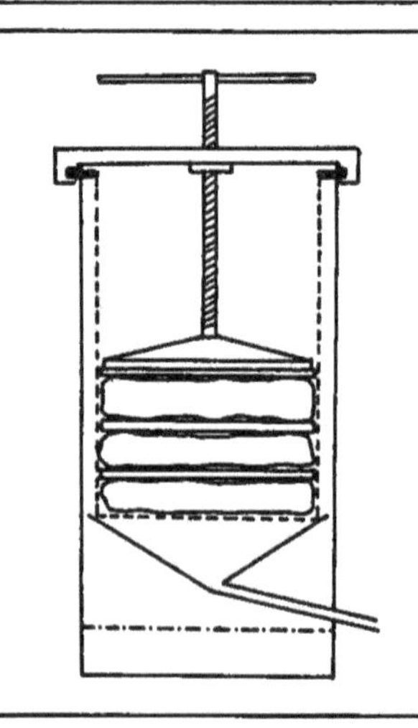

STEAM BEESWAX PRESS

The type illustrated consists of a steamer fitted with a screw plunger. Steam generated from water at the bottom of the container reaches the combs wrapped in canvas bags through the perforated basket into which they are placed; the bags are separated by wooden boards. The cross-arm is locked in position so that the combs are under pressure, and the melted wax runs through the basket and out of the tube. When the wax flow ceases, further pressure is applied. It is necessary to turn the screw back and shake up the bags before renewing the pressure, to extract all the wax, which should be done in 2 or 3 operations, leaving only 1 to 3 per cent dross behind.

Several slightly different models are supplied, one fairly similar to the drawing by:

STEFAN PUFF GmbH
Neuholdaugasse, 8011 Graz
AUSTRIA

MOUNTAIN GREY BEESWAX EXTRACTOR AND CLARIFIER

This appliance gives much cleaner wax and is very satisfactory in operation. It can be used for clean combs or dirty wax, but will not extract all the wax from cocoons or pollen of brood combs. It melts the wax in water and strains it through a coarse cloth covering the top of the container.

The two-gallon steel container has a long funnel or filler tube leading from the bottom, the top of the funnel being well above the top of the container itself. Round the outside of the container is a collecting channel for the wax, that slopes to a spout. A straining cloth is fitted over the top of the container and is held in position by means of a wire clip. The container is one-third filled with clean water, either rain water or tap water to which a little vinegar has been added (rather less than 1 per cent). The water is heated on a stove and the wax, previously soaked in water, is put into the container and the whole stirred until the wax is melted. Wax can be added until the surface comes to within 5-6cm from the top. When all the wax is completely melted, the appliance is taken off the stove and the straining cloth secured in position with the wire clip.

The wax floats to the top of the water, and more water is added through the funnel, which exerts enough pressure to force the wax through the cloth and into the collecting channel.

This extractor is obtainable from some British suppliers including:

E.H. THORNE (BEEHIVES) LTD.
Beehive Works, Wragby
Lincoln LN3 5LA
U.K.

PROPOLIS COLLECTOR

Bees gather a sticky resin known as propolis, from certain trees that produce it. They use it to close up gaps in their hive, or to reduce the hive entrance.

Propolis contains a number of antibiotic substances, derived from the plants on which the bees find it. Its use in the pharmaceutical industries of some countries is accepted. The demand for propolis, and therefore the price it fetches, vary greatly from year to year, so enquiries should be made before embarking on a programme to produce it. It is also wise to deal with an established trader: propolis is rather new as a commercial product, and therefore attracts new traders, some of whom are unable to continue. On the other hand (unlike pollen) it is relatively stable in storage. If the demand for it increases, it could provide a useful additional source of income.

The principles of harvesting are simple. A flat sheet with slits (of a width that bees will close with propolis) is inserted at the top or at the side of the hive (where the bees will regard it as the outside wall of their nest). When the slits are well closed up with propolis the sheet is removed and placed in a deep freeze; the propolis is subsequently released from the sheet by shattering.

The only firm known to market a propolis collector for placing above the top hive super is:

HUNGARONEKTÁR
Budapest 1054
Garibaldi u.2
HUNGARY

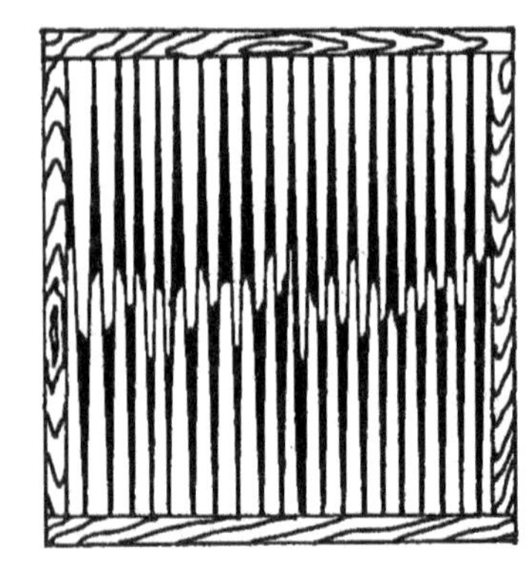

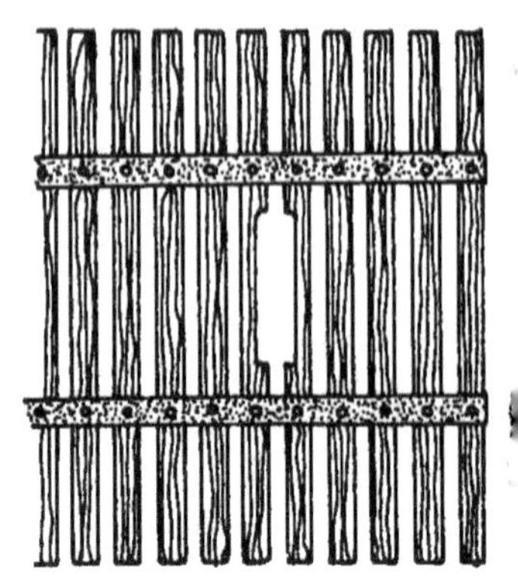

www.ingramcontent.com/pod-product-compliance
Lightning Source LLC
Jackson TN
JSHW050346140125
77033JS00020B/614
* 9 7 8 0 9 4 6 6 8 8 8 8 3 *